Green Energy and Technology

Climate change, environmental impact and the limited natural resources urge scientific research and novel technical solutions. The monograph series Green Energy and Technology serves as a publishing platform for scientific and technological approaches to "green"—i.e. environmentally friendly and sustainable—technologies. While a focus lies on energy and power supply, it also covers "green" solutions in industrial engineering and engineering design. Green Energy and Technology addresses researchers, advanced students, technical consultants as well as decision makers in industries and politics. Hence, the level of presentation spans from instructional to highly technical.

Indexed in Scopus.

Indexed in Ei Compendex.

Mohammed Al-Breiki · Yusuf Bicer

Sustainable Energy Carriers for Energy Storage and Transport

Exploring Advanced Solutions for a Green Future

Mohammed Al-Breiki
Division of Sustainable Development
College of Science and Engineering
Hamad bin Khalifa University
Doha, Qatar

Yusuf Bicer
Division of Sustainable Development
College of Science and Engineering
Hamad bin Khalifa University
Doha, Qatar

ISSN 1865-3529   ISSN 1865-3537 (electronic)
Green Energy and Technology
ISBN 978-3-031-91614-4   ISBN 978-3-031-91615-1 (eBook)
https://doi.org/10.1007/978-3-031-91615-1

Open Access funding provided by the Qatar National Library

© The Editor(s) (if applicable) and The Author(s) 2025. This book is an open access publication.

Open Access This book is licensed under the terms of the Creative Commons Attribution 4.0 International License (http://creativecommons.org/licenses/by/4.0/), which permits use, sharing, adaptation, distribution and reproduction in any medium or format, as long as you give appropriate credit to the original author(s) and the source, provide a link to the Creative Commons license and indicate if changes were made.
The images or other third party material in this book are included in the book's Creative Commons license, unless indicated otherwise in a credit line to the material. If material is not included in the book's Creative Commons license and your intended use is not permitted by statutory regulation or exceeds the permitted use, you will need to obtain permission directly from the copyright holder.
The use of general descriptive names, registered names, trademarks, service marks, etc. in this publication does not imply, even in the absence of a specific statement, that such names are exempt from the relevant protective laws and regulations and therefore free for general use.
The publisher, the authors and the editors are safe to assume that the advice and information in this book are believed to be true and accurate at the date of publication. Neither the publisher nor the authors or the editors give a warranty, expressed or implied, with respect to the material contained herein or for any errors or omissions that may have been made. The publisher remains neutral with regard to jurisdictional claims in published maps and institutional affiliations.

This Springer imprint is published by the registered company Springer Nature Switzerland AG
The registered company address is: Gewerbestrasse 11, 6330 Cham, Switzerland

If disposing of this product, please recycle the paper.

Preface

The search for more efficient, cleaner energy solutions is at the forefront of global discourse, and the urgency of climate change has become impossible to ignore. This book is an effort to unify the numerous aspects of sustainable energy carriers, including the logistical challenges of storage and transportation as well as the fundamental production processes. Our objective is to illuminate the role of each energy selected carrier in the changing energy landscape by conducting a comprehensive examination of natural gas, hydrogen, ammonia, methanol, dimethyl ether (DME), and formic acid.

This book is divided into 9 main chapters, with a concluding tenth chapter. Chapter 1, which commences the book, examines the broader context of climate change, mitigation strategies, renewable energy, various forms of energy storage, and the significance of transport infrastructure. This introductory chapter also introduces the concept of boil-off gas phenomena, demonstrating the potential impact of these losses on both economic and environmental outcomes.

The discussion expands upon this foundation by providing a comprehensive examination of each energy carrier. Chapter 2 concentrates on natural gas, focusing on its diverse forms—liquefied natural gas (LNG), compressed natural gas (CNG), and synthetic methane—and exploring the complexities of production, storage, and transportation.

In Chap. 3, hydrogen is analyzed in terms of its numerous production pathways, storage methods, and transportation solutions, emphasizing its potential as a critical component of a sustainable future.

Chapter 4 focuses on ammonia, exploring the challenges associated with its storage and transport, as well as traditional and emerging production techniques.

The narrative continues in Chap. 5, which provides a comprehensive examination of methanol. The production routes, which include fossil-based, biomass-based, and CO_2-based processes, are analyzed in conjunction with the intricacies of their storage and transportation systems.

The synthesis of dimethyl ether (DME) from methanol and syngas, as well as innovative production methods that employ captured CO_2 and green hydrogen, is the focus of Chap. 6.

Formic acid is the subject of Chap. 7, which addresses novel production methods, diverse storage strategies, and transportation considerations.

Chapter 8 provides a comparative assessment of these energy carriers, evaluating each option from the technical, economic, and life-cycle perspectives. This chapter is intended to assist readers in comprehending the trade-offs associated with various carriers, as evidenced by case studies that depict real-world scenarios and practical applications.

Lastly, Chap. 9 examines the future outlooks, highlighting emerging trends in sustainable energy systems, including decentralized and distributed energy models, advanced energy storage technologies, digitalization, and smart grids. It also addresses the increasing significance of carbon capture, utilization, and storage (CCUS) in the development of sustainable fuels, as well as policy frameworks and market trends.

Our objective in compiling this book is to offer a resource that is actionable, concise, and comprehensible for students, researchers, and professionals who are involved in the development of sustainable energy systems. We aspire to encourage further research and informed decision-making by exploring the scientific, engineering, and economic aspects of these energy carriers. May this text function as a foundational reference and a catalyst for innovative ideas in the ongoing pursuit of a more sustainable energy future.

Doha, Qatar
Mohammed Al-Breiki
Yusuf Bicer

Acknowledgements

This book is a testament to the collective mentorship, encouragement, and dedication of all those who have accompanied us.

We would like to commence by expressing our gratitude to Hamad Bin Khalifa University, Qatar Foundation, for providing the necessary infrastructure and support for our academic research during which the case studies included in this book were completed in our previous published research. All present and past members of SunReis research group are acknowledged for a peaceful and joyful research ambience.

We acknowledge Qatar National Library's support for providing this book's open access funding.

We are profoundly grateful to our wives, Souad Aqeel and Elif Derya Bicer, for their unwavering support, patience, and motivation, which served as the emotional foundation that sustained us during the extended periods of writing this book. This work has been a consistent source of inspiration due to their conviction regarding its significance.

We are also deeply grateful to our entire family (fathers: Abdullah and Mehmet; mothers: Sheikah and Sema; siblings: Amal, Saeed, Bader, Hessa, Noora, Aseela, Lulua, Laila, Haya, Al-Anood, Ali, Betul, Zeynep; children: Erva, Zumra, and Ahmet) for their unwavering love and support.

Contents

1	**Introduction**		1
	1.1 Climate Change and Mitigation Measures		1
	1.2 Renewable Energy		3
	1.3 Energy Storage		5
		1.3.1 Mechanical Storage	6
		1.3.2 Thermal Storage	7
		1.3.3 Magnetic/Electromagnetic Storage	8
		1.3.4 Electrochemical Storage	9
		1.3.5 Chemical Storage	10
	1.4 Energy Transport		11
		1.4.1 Electrical Transmission Networks	13
		1.4.2 Pipelines for Liquid and Gaseous Fuels	13
		1.4.3 Maritime and Inland Waterway Transport	14
		1.4.4 Road and Rail Transportation	14
		1.4.5 Emerging and Alternative Systems	15
	1.5 Sustainable Energy Carriers		16
		1.5.1 Electricity as an Energy Carrier	16
		1.5.2 Hydrogen as an Energy Carrier	17
		1.5.3 Ammonia as an Energy Carrier	17
		1.5.4 Biofuels as Energy Carriers	18
		1.5.5 Synthetic Fuels and Other Energy Carriers	18
	1.6 Boil-Off Gas Phenomena		19
		1.6.1 Boil-Off Gas in LNG	20
		1.6.2 Boil-Off Gas in Liquid Hydrogen	20
		1.6.3 Boil-Off Gas in Other Liquefied Energy Carriers	21
		1.6.4 Strategies for Managing Boil-Off Gas	21
	References		22
2	**Natural Gas**		25
	2.1 Liquefied Natural Gas		25
		2.1.1 LNG Production	25

		2.1.2	LNG Storage	27
		2.1.3	LNG Transport	28
	2.2	Compressed Natural Gas		30
		2.2.1	CNG Production	30
		2.2.2	CNG Storage	31
		2.2.3	CNG Transport	32
	2.3	Synthetic Natural Gas		33
		2.3.1	SNG Production	33
		2.3.2	SNG Storage	35
		2.3.3	SNG Transport	36
	Reference			37
3	**Hydrogen**			**39**
	3.1	Hydrogen Production		39
		3.1.1	Steam Methane Reforming	40
		3.1.2	Methane Pyrolysis	41
		3.1.3	Partial Oxidation of Hydrocarbons	42
		3.1.4	Plasma Reforming of Hydrocarbons	43
		3.1.5	Coal Gasification	44
		3.1.6	Electrolysis of Water	46
		3.1.7	Photobiological Processes	46
		3.1.8	Photoelectrochemical Water Splitting	49
		3.1.9	Thermochemical Water Splitting	49
		3.1.10	Biomass Gasification	50
	3.2	Hydrogen Storage		51
		3.2.1	Physical Storage	52
		3.2.2	Chemical Storage	53
		3.2.3	Materials-Based Storage	55
	3.3	Hydrogen Transport		56
	References			58
4	**Ammonia**			**61**
	4.1	Ammonia Production		61
		4.1.1	The Haber-Bosch Process	61
		4.1.2	Electrochemical Ammonia Synthesis	63
		4.1.3	Plasma-Based Ammonia Synthesis	64
		4.1.4	Biological Ammonia Synthesis	68
		4.1.5	Ammonia Synthesis via Methane Pyrolysis	69
		4.1.6	Renewable Ammonia Synthesis	70
	4.2	Ammonia Storage		70
		4.2.1	Pressurized Storage	71
		4.2.2	Refrigerated Storage	72
		4.2.3	Solid-State Storage	72
		4.2.4	Liquid Ammonia	73
		4.2.5	Absorption in Porous Materials	74

	4.3	Ammonia Transport	75
		4.3.1 Pressurized Tankers and Barges	75
		4.3.2 Pressurized Rail Tank Cars and Tank Trucks	77
		4.3.3 Pipelines	78
	References		80
5	**Methanol**		**83**
	5.1	Methanol Production	83
		5.1.1 Steam Methane Reforming	83
		5.1.2 Coal Gasification	86
		5.1.3 Biomass Gasification	87
		5.1.4 CO_2 Hydrogenation	90
		5.1.5 Electrochemical CO_2 Reduction	92
	5.2	Methanol Storage	93
		5.2.1 Physical Storage	94
		5.2.2 Chemical Storage	96
		5.2.3 Materials-Based Storage	96
	5.3	Methanol Transport	97
		5.3.1 Maritime Transport	97
		5.3.2 Land Transport	98
		5.3.3 Pipelines	100
	References		103
6	**Dimethyl Ether**		**105**
	6.1	DME Production	105
		6.1.1 Methanol Dehydration	106
		6.1.2 Direct DME Synthesis from Syngas	107
		6.1.3 Biomass Gasification	108
		6.1.4 DME Production from Captured CO_2 and Green Hydrogen	110
	6.2	DME Storage	111
		6.2.1 Physical Storage	111
		6.2.2 Chemical Storage	114
		6.2.3 Material-Based Storage	116
	6.3	DME Transport	118
		6.3.1 Maritime Transport	118
		6.3.2 Land Transport	119
		6.3.3 Pipelines	119
	References		121
7	**Formic Acid**		**123**
	7.1	Formic Acid Production	123
		7.1.1 Methyl Formate Hydrolysis	123
		7.1.2 Biomass Oxidation	125
		7.1.3 Electrochemical Reduction of CO_2	125
		7.1.4 Hydrolysis of Formamide	127

	7.2	Formic Acid Storage	128
		7.2.1 Physical Storage	128
		7.2.2 Chemical Storage	130
		7.2.3 Material-Based Storage	132
	7.3	Formic Acid Transport	134
		7.3.1 Maritime Transport	134
		7.3.2 Land Transport	135
		7.3.3 Pipelines	135
	References		137
8	**Overall Comparison of Energy Carriers**		**139**
	8.1	Technical Assessment of Liquefied Energy Carriers	139
		8.1.1 Energy Requirements for Production	139
		8.1.2 Energy Requirements for Storage	142
		8.1.3 Energy Requirements for Transportation	144
		8.1.4 Comparison and Analysis	146
		8.1.5 Case Study 1: Technical Assessment of Energy Carriers' Entire Supply Chain	147
	8.2	Economic Assessment	161
		8.2.1 Costs Associated with Production	161
		8.2.2 Costs Associated with Storage	163
		8.2.3 Costs Associated with Transportation	165
		8.2.4 Comparison and Analysis	167
		8.2.5 Case Study 2: Economic Assessment of Energy Carriers' Entire Supply Chain	169
	8.3	Life Cycle Assessment	186
		8.3.1 Emissions Associated with Production	186
		8.3.2 Emissions Associated with Storage	189
		8.3.3 Emissions Associated with Transportation	191
		8.3.4 Comparison and Analysis	194
		8.3.5 Case Study 3: Life Cycle Assessment of Energy Carriers' Entire Supply Chain	195
	References		221
9	**Future Directions**		**227**
	9.1	Future Energy Systems	227
		9.1.1 Decentralized and Distributed Energy Systems	227
		9.1.2 Advanced Energy Storage Technologies	228
		9.1.3 Digitalization and Smart Grids	229
		9.1.4 Renewable Energy Integration and Hybrid Systems	230
		9.1.5 Carbon Capture, Utilization, and Storage	231
	9.2	Future Sustainable Fuel Market	232
		9.2.1 Market Trends for Hydrogen and Ammonia	232
		9.2.2 Policy and Regulatory Drivers	233
		9.2.3 Investment and Financing in Sustainable Fuels	234
		9.2.4 Future Market Challenges	234

9.3	Utilization of Sustainable Energy Carriers		235
	9.3.1	Hydrogen as a Key Player in Future Energy Systems	235
	9.3.2	Ammonia as an Energy Carrier and Fuel	236
	9.3.3	Methanol in the Transport and Power Sectors	237
	9.3.4	DME in Transportation and Power Generation	238
	9.3.5	Integration of Sustainable Carriers into Existing Energy Systems	239
9.4	Concluding Remarks		240
References			241

Acronyms

BOG	Boil-off gas
CIS	Commonwealth of Independent States
DME	Dimethyl Ether
eq	Equivalent
GHG	Greenhouse gas
GREET	The Greenhouse Gases, Regulated Emissions, and Energy Use in Transportation
GTL	Gas-to-liquid
GWP	Global warming potential
HFO	Heavy fuel oil
ICS	International Chamber of Shipping
IMO	International Maritime Organization
$L-H_2$	Liquid hydrogen
LNG	Liquefied natural gas
$L-NH_3$	Liquid ammonia
LPG	Liquid petroleum gas
MIC	Mesaieed Industrial City
MJ	Megajoule
MT	Million ton

nmi	Numerical mile
PV	Photovoltaic
tcf	Trillion cubic feet
TJ	Terajoule

Nomenclature

C_p	Isobaric specific heat capacity (kJ/kg °C)
D	Inner diameter (m)
E	Specific energy (MJ/kg)
HC	Total heat capacity of the tank (kJ/°C)
K	Thermal conductivity (W/m °C)
L	Length of the transfer pipe (m)
LHV	Lower heating value (kJ/kg)
M_H	Mass of the heel (kg)
Q	Heat transfer rate (W)
t	The thickness of insulation (in)
U	Overall heat transfer coefficient (W/m² K)
Δh_{vap}	The heat of vaporization (kJ/kg)

Chapter 1
Introduction

1.1 Climate Change and Mitigation Measures

The increasing frequency and severity of climate-related events is one of the most urgent challenges of the twenty-first century. The inexorable rise in global temperatures, caused by the accumulation of greenhouse gases (GHGs) in the Earth's atmosphere, has initiated a series of environmental, social, and economic disruptions. According to the Intergovernmental Panel on Climate Change (IPCC), the Earth has already experienced a temperature increase of about 1.1 °C compared to pre-industrial levels. This has resulted in significant impacts such as increased occurrence and intensity of extreme weather events, higher sea levels, and loss of biodiversity. The impacts of climate change are not uniformly distributed worldwide. Certain regions, like the Arctic, are experiencing warming at a rate that is more than double the global average. This accelerated warming has resulted in rapid ice melting and significant disruptions to ecosystems and local communities [1].

The scientific consensus is unequivocal: unless significant and prompt measures are taken to reduce GHG emissions, the world is projected to surpass crucial thresholds, resulting in catastrophic and potentially irreversible consequences. The IPCC's Sixth Assessment Report states that in order to restrict global warming to 1.5° C, there needs to be a reduction of approximately 45% in global CO_2 emissions from 2010 levels by 2030, with the aim of achieving net-zero emissions by around 2050 [1]. Surpassing this limit could activate tipping points in the Earth's climate system, such as the disintegration of polar ice sheets or the disturbance of significant ocean currents, which could intensify the impacts of climate change and hinder its reversal.

Implementing mitigation measures is crucial in the worldwide endeavor to address and counteract climate change. These measures are intended to mitigate the release of GHGs, thus restricting the magnitude of climate change and its corresponding consequences. The Paris Agreement, ratified in 2015, represents a major achievement in global climate policy by aiming to restrict the increase in average global temperature to less than 2 °C, and ideally to 1.5 °C, compared to pre-industrial levels.

In order to reach this goal, a radical change is required in the methods by which societies generate and use energy, as energy production and consumption contribute significantly to worldwide greenhouse gas emissions [2].

The energy sector plays a significant role in both causing climate change and providing opportunities for reducing its impact. The combustion of fossil fuels for electricity, heat, and transportation is the primary cause of the majority of GHG emissions worldwide, specifically CO_2. The shift from dependence on fossil fuels to low-carbon and renewable energy sources is crucial for mitigating emissions. Nevertheless, achieving decarbonization in the energy sector necessitates progress in energy storage and transport technologies to guarantee the sustainable production of clean energy and its efficient integration into the grid for delivery to consumers [3].

Renewable energy technologies, including wind, solar, and hydropower, are highly efficient means of addressing climate change. They provide the advantages of both decreasing reliance on fossil fuels and reducing GHG emissions. Nonetheless, the irregularity of renewable energy sources poses a notable obstacle for energy systems that heavily depend on fossil fuels for continuous power supply. Energy storage and transport solutions are crucial in this context. Technological advancements such as batteries, hydrogen storage, and advanced grid systems enable a consistent provision of renewable energy, even in the face of variable generation caused by weather conditions [4].

Furthermore, alongside the shift towards renewable energy, mitigation strategies encompass improving energy efficiency, advocating for sustainable land use practices, and advancing technologies for carbon capture, utilization, and storage (CCUS). Enhancements in energy efficiency can greatly diminish the need for energy, resulting in a substantial decrease in emissions. Energy efficiency improvements alone has the potential to contribute over 40% of the necessary emissions reductions to achieve global climate objectives [3]. In addition, the implementation of sustainable land use practices, such as reforestation and sustainable agriculture, can augment the natural capacity for CO_2 absorption, thereby facilitating a decrease in atmospheric carbon levels [5].

CCUS technologies are an essential part of efforts to mitigate climate change. CCUS entails the capture of CO_2 emissions directly from their origin, such as power plants or industrial facilities, and subsequent underground storage or utilization in industrial operations. Although CCUS cannot replace the need to reduce emissions directly, it is seen as an essential addition to other methods of reducing emissions, especially in sectors like heavy industry and aviation, where it is challenging to completely eliminate emissions [1].

Furthermore, the significance of policy frameworks and international cooperation cannot be emphasized enough. Efficient climate action necessitates synchronized endeavors among nations, sectors, and levels of governance. Nationally Determined Contributions (NDCs), which are submitted by countries in accordance with the Paris Agreement, delineate the national strategies and objectives for mitigating greenhouse gas emissions. Nevertheless, the effectiveness of these measures relies on how well they are put into practice and the capacity of nations to strengthen their obligations as time goes on. The Emissions Gap Report by the UN Environment Programme emphasizes that the current NDCs are inadequate to achieve the targets set by the Paris Agreement. It suggests that more ambitious actions are required, especially from the world's major polluters [6].

To summarize, climate change is a complex problem that requires a comprehensive strategy for reducing its impact. Adopting a low-carbon economy, facilitated by advancements in energy storage and transportation, is crucial for attaining the climate objectives established by the global community. This book will examine the significant role that sustainable energy carriers play in the transition to a decarbonized world. These energy carriers provide solutions to the challenges of storing and transporting energy in an environmentally friendly manner. The following sections will explore these topics, starting with an analysis of renewable energy as a fundamental aspect of climate change mitigation initiatives.

1.2 Renewable Energy

Renewable energy is at the forefront of worldwide endeavors to address climate change, providing a feasible approach to substantially decrease GHG emissions while satisfying the increasing energy needs. With the ongoing growth of the global population and the rise in energy consumption due to economic development, there is an urgent requirement for sustainable and low-carbon energy sources. Renewable energy, sourced from ongoing natural processes including sunshine, wind, rain, tides, waves, and geothermal heat, is a fundamental element of this shift.

The main benefit of renewable energy lies in its capacity to generate electricity and other forms of energy without significant carbon emissions. Renewable energy sources, unlike fossil fuels, produce power without the need for combustion, resulting in a minimal carbon footprint as they do not create substantial volumes of CO_2 when utilized. This attribute is essential in the battle against climate change, as the energy sector bears significant responsibility for around 75 % of worldwide GHG emissions, predominantly stemming from the combustion of coal, oil, and natural gas [3].

Solar energy is a plentiful and rapidly expanding type of renewable energy. Photovoltaic (PV) technology, which directly transforms sunlight into electricity, has shown notable progress in recent decades, resulting in considerable reductions in cost and gains in efficiency. According to the International Renewable Energy Agency (IRENA), the price of electricity generated from large-scale solar PV systems decreased by 85 % from 2010 to 2020. This significant reduction in cost has made solar PV one of the most economically viable energy options in many regions worldwide [7]. Furthermore, alongside PV technology, concentrated solar power (CSP) systems are being actively researched and developed. These systems utilize mirrors or lenses to concentrate sunlight into a tiny area, generating heat. The aim is to harness this technology to generate large-scale, dispatchable power.

Wind energy is a significant contributor to the renewable energy sector. Wind farms both on land and at sea have become more common, especially in areas with abundant and reliable wind conditions. Recent advancements in turbine technology have resulted in the development of larger and more efficient devices that are capable of harnessing a greater amount of energy from the wind. In 2020, the global wind power capacity grew by 93 GW, as reported by the Global Wind Energy Council (GWEC). This resulted in a total installed capacity of over 743 GW [8]. Offshore wind power is becoming increasingly popular because it has greater capacity factors and allows for the use of larger turbines that cannot be used on land.

Hydropower, which is the primary source of renewable electricity on a global scale, continues to have a substantial impact on the energy composition. It offers a dependable and adaptable power source that can meet both constant energy demands and ensure grid stability by rapidly adjusting to shifting electrical requirements. Although large-scale hydropower projects can cause substantial environmental and social consequences, such as disrupting ecosystems and displacing populations, there are ongoing initiatives to create smaller run-of-river systems that have a reduced environmental impact [9].

Geothermal energy, which utilizes heat from the Earth's interior, presents a promising opportunity for generating renewable power, especially in areas abundant in geothermal resources. Geothermal power plants offer a consistent and dependable energy supply while producing minimum pollutants. Geothermal energy differs from solar and wind energy in that it is constant and consistent, available at all times. This characteristic makes geothermal energy a crucial element of a well-rounded and robust energy system [10].

Bioenergy is a flexible type of renewable energy that can be utilized for power, heat, and transportation fuels. It is obtained from organic materials such as agricultural wastes, forestry by-products, and organic waste. Bioenergy possesses the distinct benefit of being capable of supplying energy in forms that can directly replace fossil fuels, such as liquid biofuels for transportation. Nevertheless, the viability of bioenergy hinges on the origin of the biomass and the techniques employed in its production, since unsustainable practices can result in deforestation, depletion of biodiversity, and conflicts with food production [11].

Marine energy, encompassing tidal, wave, and ocean thermal energy conversion (OTEC), is still in its early stages of development in comparison to other renewable technologies. Nevertheless, it possesses considerable promise, especially in coastal and Island areas. Tidal energy utilizes the consistent ebb and flow of ocean tides, while wave energy collects the power generated by surface waves. OTEC harnesses the thermal gradient between the warmer surface water and the colder deep water in order to produce energy. As these technologies advance, they have the potential to play a significant role in the renewable energy portfolio [12].

The incorporation of renewable energy into the current energy infrastructure poses both difficulties and prospects. An important obstacle lies in the fluctuation of renewable energy sources, including solar and wind. In order to tackle this issue, energy storage technologies, such as batteries and pumped hydrostorage, are being created to store surplus energy during periods of overproduction and release it during periods of underproduction. In addition, advancements in grid management and smart grid technology are facilitating the equilibrium of supply and demand, hence enabling a greater integration of renewable energy sources in the energy composition [3].

The fast spread of renewable energy has been greatly facilitated by policy assistance. Various governments worldwide have adopted a variety of measures and incentives, such as feed-in tariffs, renewable portfolio standards, tax credits, and subsidies, to promote the use of renewable energy technologies. The European Union's Green Deal seeks to achieve climate neutrality for Europe by 2050, with a particular emphasis on the integration of renewable energy sources [13].

Overall, renewable energy is an essential element of the worldwide plan to reduce the impact of climate change. In order to reach worldwide climate targets, it is crucial to continue developing and implementing renewable energy technologies while also focusing on energy storage and grid management advances, all of which must be backed by suitable laws. These efforts are necessary for attaining deep decarbonization. This book will delve into the function of renewable energy, highlighting its significance not only in generating electricity but also in producing and using sustainable energy carriers. Subsequent sections will provide detailed discussions on this topic.

1.3 Energy Storage

Energy storage plays a crucial role in facilitating the shift towards a low-carbon energy system by actively tackling the major obstacle related to renewable energy sources: their inherent variability. The increasing integration of renewable energy sources, such as solar and wind, into the global energy mix necessitates the development of efficient, scalable, and cost-effective energy storage solutions. Energy storage technologies enable the acquisition and retention of energy generated during times of excess production, which can subsequently be deployed when demand surpasses supply or when renewable generation is insufficient. This capability is crucial for preserving the stability of the power grid, guaranteeing a dependable energy provision, and enabling the incorporation of renewable energy into the power generation system.

The energy storage technology landscape is characterized by its diversity, including a wide array of systems that differ in terms of their capacity, discharge duration, response time, and applications. In general, these technologies can be classified into electrochemical, mechanical, thermal, chemical storage systems, and magnetic, each possessing distinct benefits and difficulties. Figure 1.1 illustrates a hierarchical classification of energy storage technologies, grouping them into five principal categories: mechanical, thermal, magnetic/electromagnetic, electrochemical, and chemical. Each main category then branches into various subcategories—for example, mechanical energy storage is shown to include pumped hydro, compressed air, and flywheels, while chemical energy storage comprises hydrogen, ammonia, synthetic natural gas, methanol, formic acid, and dimethyl ether. This arrangement

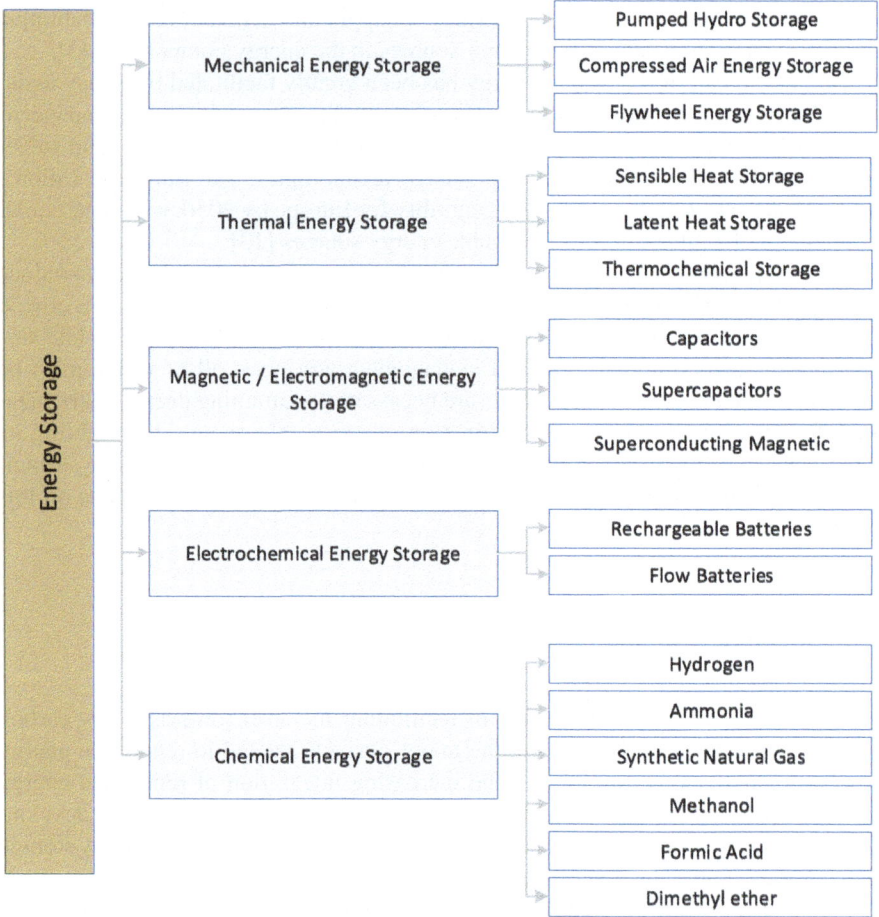

Fig. 1.1 Overview of energy storage technologies, illustrating the main categories and subcategories commonly used for large- and small-scale applications

highlights the broad range of solutions under development or in use to address different application needs, storage durations, and performance requirements.

1.3.1 Mechanical Storage

Mechanical energy storage systems are fundamental to large-scale energy management, as they convert and store energy in physical forms. Principal technologies in this field encompass pumped hydrostorage (PHS), compressed air energy storage (CAES), and flywheel energy storage.

1.3 Energy Storage

PHS is the most advanced and extensively adopted technology. It operates by transferring water from a lower reservoir to an upper reservoir during times of reduced demand. When electricity is required, water circulates back through turbines to produce power. This method constitutes more than 90% of the worldwide installed storage capacity [14]. Its capacity to provide substantial energy over prolonged durations, coupled with additional grid services like frequency regulation and reserve power, highlights its essential function in contemporary energy systems.

CAES functions on a comparable principle, substituting water with air. During off-peak periods, energy is utilized to compress air, which is subsequently stored in subterranean caverns or specialized tanks. During periods of peak demand, compressed air is discharged to activate turbines and generate electricity. While CAES is less common than PHS, it provides similar advantages regarding scale and discharge duration. Its implementation, however, may be restricted by geological limitations and frequently encounters efficiency issues due to energy losses during the compression and expansion phases.

Flywheel energy storage employs an alternative method by conserving energy in the form of rotational kinetic energy. This system accelerates a rotor to high speeds utilizing electrical energy, preserving the kinetic energy until required. When energy is needed, the rotor's power is rapidly transformed back into electricity. Although flywheels typically possess lower energy density than PHS and CAES, they are superior in applications necessitating swift power delivery and elevated power density, including frequency stabilization and short-term backup power. Their capacity for rapid response renders them an appealing choice for mitigating power fluctuations and enhancing grid stability [15].

Collectively, these mechanical energy storage technologies provide a varied array of solutions, each possessing distinct advantages tailored to different applications, ranging from long-duration energy storage to swift grid support.

1.3.2 Thermal Storage

Thermal energy storage (TES) systems capture and retain energy as heat, offering essential support for applications including space heating, industrial processes, and electricity generation. TES technologies can be categorized into three distinct classifications. (i) Sensible heat storage represents the predominant variant of TES [16]. Energy is accumulated by elevating the temperature of a substance—such as water, molten salts, or rocks—during times of low demand, and subsequently retrieved by cooling the substance when energy is required. An exemplary instance is molten salt storage, extensively utilized in concentrated solar power (CSP) facilities. In these systems, solar energy heats the salt during the day, which is subsequently utilized to generate steam for electricity production at night or during overcast conditions. (ii) Latent heat storage utilizes phase change materials (PCMs) for energy storage. During a phase transition, such as from solid to liquid, substances like paraffin wax or salt hydrate absorb substantial energy without an increase in temperature. The reverse process (solidification) discharges this stored energy at a relatively stable temperature, rendering latent heat storage an effective method for managing thermal loads with minimal temperature variation. (iii) In thermochemical storage, energy is conserved via reversible chemical reactions. Rather than merely heating a material, these systems employ endothermic reactions (which absorb energy) to accumulate heat and exothermic reactions (which release energy) to retrieve it. This method provides elevated energy densities and prolonged storage capabilities, as the energy is retained within the chemical bonds until the reaction is inverted. Despite being in active development, thermochemical storage is regarded as a promising solution for applications necessitating prolonged storage durations and efficient energy retrieval.

Every TES modality presents distinct advantages and challenges. Sensible and latent heat storage have been extensively utilized in industrial applications and district heating networks; however, their integration into power grids for electricity generation remains intricate. The primary challenge is the efficient transformation of stored thermal energy into electrical energy, requiring sophisticated and effective heat-to-power technologies.

1.3.3 Magnetic/Electromagnetic Storage

Magnetic and electromagnetic energy storage systems harness and discharge energy by manipulating electric or magnetic fields, rendering them suitable for applications necessitating rapid charge and discharge cycles alongside high-power outputs. These technologies are crucial for grid frequency regulation, immediate backup power, and scenarios where swift response and power quality are critical.

Conventional capacitors represent the fundamental type of electromagnetic storage. They comprise two conductive plates divided by a dielectric substance. Upon the application of voltage, an electric field is generated as charges amass on the plates, facilitating the rapid storage of energy. Their advantages encompass swift response times and extended lifespans, as they are not subject to the chemical degradation that impacts batteries. Nonetheless, the energy density of traditional capacitors is comparatively low, restricting their application to brief, high-power scenarios.

Supercapacitors serve as an intermediary between traditional capacitors and batteries. They employ sophisticated electrode materials with remarkably high surface areas to substantially enhance energy storage capacity while simultaneously providing rapid power bursts. Supercapacitors can withstand a substantial number of charge–discharge cycles without performance deterioration, rendering them suitable for applications such as regenerative braking in vehicles, industrial power stabilization, and buffering intermittent renewable energy sources. Their enhanced energy density compared to conventional capacitors allows them to function in applications where both power and energy storage are essential.

Superconducting magnetic energy storage systems retain energy within the magnetic field produced by a direct current flowing through a superconducting coil, which is kept at cryogenic temperatures. Superconductors demonstrate minimal electrical resistance, resulting in remarkably high efficiencies with minimal energy loss over time. SMES units can provide nearly instantaneous power, essential for applications such as grid stabilization and power quality management. Notwithstanding these benefits, the necessity for intricate, energy-demanding cooling systems and substantial capital expenditures has thus far constrained SMES implementations to specialized applications and research environments.

These magnetic and electromagnetic storage technologies collectively provide distinct advantages, including swift energy delivery and extended cycle life, which enhance other energy storage solutions. Their advancement and incorporation persist in improving our capacity to regulate power quality and stability in contemporary electrical grids.

1.3.4 Electrochemical Storage

Electrochemical storage encompasses more than the prevalent use of lithium-ion technology, as various alternative systems are currently being developed to meet specific application requirements and mitigate inherent limitations. Electrochemical storage, mainly lithium-ion batteries, has emerged as the prevailing process for energy storage, propelled by advancements in battery technology and the swift decline in manufacturing costs. Based on their high energy density, efficiency, and extended cycle life, lithium-ion batteries are well-suited for a diverse array of applications, including portable electronics, electric vehicles (EVs), and grid-scale storage [14]. The cost of lithium-ion batteries has experienced a reduction of over 85% during the last ten years, resulting in an average price of $139 per kilowatt-hour (kWh) in 2023. This downward trend is projected to persist as production increases and technological advancements are achieved [17].

Nevertheless, lithium-ion batteries are not exempt from their constraints. Manufacturing these batteries necessitates substantial quantities of essential raw materials, including lithium, cobalt, and nickel, which present supply chain vulnerabilities and environmental obstacles [18]. Moreover, the energy density of lithium-ion batteries, although excellent, may not meet the requirements of certain large-scale grid applications. This has prompted the investigation of other battery technologies, including solid-state batteries, flow batteries, and sodium-ion batteries [19].

Flow batteries store energy in liquid electrolytes housed in external tanks, which circulate through an electrochemical reactor during charging and discharging processes. This architecture separates energy capacity from power output, rendering it highly scalable and particularly suitable for grid-scale applications where prolonged storage duration is essential. Vanadium redox flow batteries are distinguished by their extended cycle life and stability, although they exhibit a lower energy density relative to lithium-ion systems. Their modular design and relative ease of maintenance render them promising candidates for the integration of substantial proportions of renewable energy into the grid.

Fuel cells embody a distinct methodology by directly converting the chemical energy of fuels, typically hydrogen, into electricity via electrochemical reactions, instead of internally storing energy. Hydrogen fuel cells provide exceptional efficiency and swift power output, emitting solely water as a byproduct. Nonetheless, their implementation encounters obstacles concerning the production, storage, and distribution of hydrogen, in addition to elevated initial costs relative to battery-based systems. Notwithstanding these challenges, fuel cells are being utilized in various sectors, including transportation—where they energize fuel cell electric vehicles—and stationary power generation, frequently augmenting battery storage in hybrid energy systems.

Collectively, these nascent technologies underscore the evolving trend of electrochemical storage, wherein the search for enhanced performance, safety, and sustainability propels ongoing innovation beyond the traditional lithium-ion framework.

1.3.5 Chemical Storage

Chemical energy storage offers a viable method for harnessing excess renewable energy and transforming it into storable, transportable fuels. This method not only mitigates seasonal variations in renewable energy production but also facilitates the decarbonization of multiple sectors. The subsequent carriers exemplify the range of chemical storage solutions.

Hydrogen continues to be a leading method for chemical energy storage. Produced via electrolysis, it can be stored and subsequently converted back to electricity using fuel cells or combustion engines. Its ability to store substantial quantities of energy for extended periods renders it particularly effective in alleviating seasonal fluctuations in renewable energy availability. Additionally, hydrogen serves as a multifaceted energy carrier, utilized in transportation, heating, and industrial processes [20].

Ammonia, synthesized from hydrogen and nitrogen through methods like Haber–Bosch, possesses a high volumetric energy density and utilizes current infrastructure for storage and transportation. Ammonia's application as a carbon-free fuel in combustion systems and specialized fuel cells establishes it as a pivotal component in industries such as shipping and power generation. The simplicity of liquefaction and the comparative stability of ammonia augment its viability as an effective energy carrier.

Synthetic natural gas, also referred to as renewable natural gas (if the source of electricity is renewable), is generated by the reaction of hydrogen with captured carbon dioxide via methanation processes. This conversion both stores renewable energy as a gas compatible with current natural gas infrastructure and recycles CO_2, thereby aiding in the reduction of greenhouse gases. The incorporation of this technology into the existing gas infrastructure provides a means for decarbonizing heating and industrial operations.

Methanol, synthesized from hydrogen and carbon dioxide, is a liquid fuel that facilitates handling, storage, and distribution more efficiently than gaseous fuels. It serves as both a direct fuel in internal combustion engines and direct methanol fuel cells, as well as a significant feedstock in the chemical industry. The liquid state of methanol and its established market presence render it a viable choice for energy storage and conversion.

Formic acid is emerging as a promising liquid hydrogen carrier, offering a safer and more convenient alternative to conventional hydrogen storage methods. It can retain hydrogen at comparatively low pressures and be catalytically decomposed to liberate hydrogen as needed. This attribute renders formic acid a compelling choice for applications necessitating on-site hydrogen production without the intricacies associated with managing gaseous hydrogen.

DME is an alternative fuel produced from methanol or synthesized directly from syngas. DME, possessing characteristics akin to liquefied petroleum gas, features a high cetane number and combusts more cleanly than traditional diesel. DME's compatibility with current fuel distribution systems and its applicability in diesel engines make it an attractive option for transportation and power generation.

Collectively, these chemical storage alternatives provide varied solutions for tackling energy storage issues. Each carrier offers distinct advantages—hydrogen's versatility, ammonia's infrastructure compatibility, and methanol's handling simplicity— contributing to a future where renewable energy can be effectively stored, transported, and employed across various applications.

1.4 Energy Transport

The techniques for energy transportation are diverse, mirroring the variety of energy carriers and the magnitude of energy requirements. In the contemporary interconnected world, the effective transfer of energy from production locations to consumption sites is essential for economic development, national security, and environmental sustainability. Energy transport systems must address various aspects, including the physical characteristics of the energy carrier, geographical distances, and infrastructural obstacles related to delivery. The design and implementation of energy transfer technologies are specialized and contingent upon context.

The primary methods of energy transportation encompass electrical transmission networks, pipelines, maritime and inland waterways, as well as road and rail systems. Each modality is specifically designed to meet distinct energy carriers and operational needs. Electrical transmission networks are designed to reduce losses in the long-distance transport of energy, utilizing high-voltage alternating current (HVAC) or high-voltage direct current (HVDC) systems. Pipelines offer a continuous and secure method for delivering liquid and gaseous fuels, including oil, natural gas, and, increasingly, hydrogen. Maritime and inland waterways facilitate the transportation of substantial quantities of energy carriers, such as crude oil and LNG, across international distances, while road and rail networks allow for the delivery of fuels to regions lacking fixed infrastructure.

This section examines the diverse methods of energy transmission, highlighting the adaptation of each method to the physical and chemical properties of its corresponding energy carrier. This discussion seeks to elucidate the challenges and opportunities present in contemporary energy logistics by analyzing the technological, economic, and environmental dimensions of these transportation networks. As global energy demands grow, solutions for energy transportation must also adapt to maintain efficiency, resilience, and sustainability among evolving trends and technology. Figure 1.2 provides a systematic classification of energy transport systems, dividing them into five primary categories: electrical transmission networks encompass high-voltage AC/DC systems and smart grids for optimized power distribution; pipeline transport includes oil, natural gas, and developing hydrogen pipelines; maritime and inland waterway transport features oil tankers, LNG carriers, and inland barges for extensive fuel shipping; road and rail transportation employs fuel tanker trucks and rail tank cars for adaptable land-based energy distribution; and emerging and alternative systems comprise decentralized energy transport such as microgrids, battery storage, and Power-to-X technologies. These transportation methods are crucial for guaranteeing the efficient and secure transfer of energy resources globally.

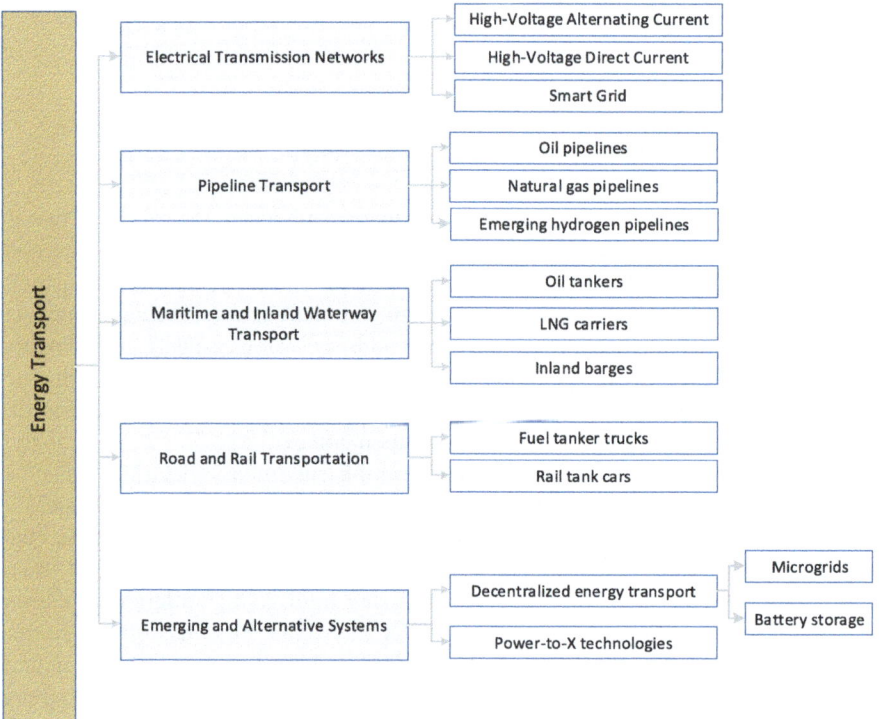

Fig. 1.2 Energy transport systems and their key components

1.4.1 Electrical Transmission Networks

Electrical transmission networks are pivotal to the contemporary energy framework, functioning as the principal conduit for electricity distribution—a multifaceted and extensively utilized secondary energy carrier—from generation locations to end-users. Due to its ability for long-distance transmission with minimal losses, especially when employing high-voltage systems, electricity has emerged as the foundation of modern energy systems [3]. These networks are primarily classified into alternating current (AC) and direct current (DC) systems, each presenting unique benefits [21]. Alternating current systems have historically been utilized for long-distance transmission primarily because of the convenience of voltage transformation through transformers. Conversely, progress in HVDC technology has resulted in reduced transmission losses and supported the integration of asynchronous grids, underscoring the dynamic advancement of electrical transmission techniques [22].

The foundational grid infrastructure is intrinsically intricate and interconnected, requiring meticulous management to guarantee stability and reliability. Modernization initiatives, particularly the integration of smart grid technologies, are essential for improving the grid's ability to support a growing proportion of renewable energy sources and decentralized power generation. Nonetheless, despite the efficiency improvements realized via high-voltage transmission, intrinsic challenges remain. Energy losses during transmission, although reduced, continue to be a concern over extensive distances. Moreover, a considerable segment of current transmission networks is deteriorating, highlighting the pressing necessity for extensive enhancements to accommodate future demand and strengthen system resilience.

1.4.2 Pipelines for Liquid and Gaseous Fuels

Pipelines serve as an exceptionally efficient means for the continuous transportation of liquid and gaseous fuels, including crude oil, natural gas, and, progressively, hydrogen. Their design minimizes handling, mitigates the risk of spillage, and offers a cost-efficient solution for transporting substantial quantities of energy carriers across extensive distances. Pipelines are designed for operational efficiency, providing continuous, high-capacity transport that consistently delivers fuels to downstream processing facilities and distribution centers. Advanced sensors and automated control systems are essential to their operation, continuously monitoring for leaks, pressure fluctuations, and other anomalies, thus improving safety and minimizing environmental risks [23].

Emerging trends in pipeline technology highlight their essential function in contemporary energy infrastructure. Amid the global focus on decarbonization, there is increasing interest in repurposing current natural gas pipelines or constructing new ones for hydrogen transportation. Nevertheless, the technical obstacles related to hydrogen, including its diminutive molecular size and elevated diffusivity, necessitate inventive engineering solutions. Furthermore, natural gas pipelines are being modified in specific areas to transport blended fuels that may include bio-derived gases, thus facilitating the incorporation of renewable energy sources into conventional energy transport systems [24].

1.4.3 Maritime and Inland Waterway Transport

Maritime and inland waterway transport constitute an essential foundation of global energy logistics, significantly facilitating the transportation of vital energy carriers, including crude oil, LNG, and refined petroleum products. This transportation mode is essential for areas with extensive river systems, where inland waterways enhance maritime routes to maintain a steady supply of energy resources. Large-scale tankers and specialized LNG carriers are integral to this system, as they are designed to improve energy density and ensure maximum safety. These vessels are engineered with sophisticated features to reduce the risks of spillage and explosion, guaranteeing the efficient transportation of substantial fuel quantities across international waters [25].

Maritime transport provides exceptional geographic flexibility, which is crucial for enabling the import and export of energy resources across continents. This flexibility is crucial for nations devoid of domestic energy reserves, as it allows them to engage with global markets and obtain essential supplies. Notwithstanding these benefits, maritime and inland waterway transport encounter considerable challenges. Environmental risks are a primary concern; the possibility of oil spills, accidental discharges, and greenhouse gas emissions highlights the ecological consequences of these transportation methods [26]. The regulatory framework for maritime transport is intricate and constantly changing, shaped by numerous geopolitical and environmental influences. These challenges require continuous improvements in vessel design, operational procedures, and international regulatory frameworks to guarantee that maritime and inland waterway transport effectively meets global energy demands safely and sustainably [27].

1.4.4 Road and Rail Transportation

Road and rail transportation are essential components of the energy supply chain, especially for regional and local distribution, while pipelines and maritime transport primarily facilitate large-scale movements. Trucks, rail tank cars, and barges are frequently utilized for the final stage of energy distribution, guaranteeing that energy carriers reach a diverse array of consumers, including those in isolated or non-pipeline-connected regions. This mode of transport is characterized by its versatility, enabling access to locations that larger, fixed infrastructures may not reach, and by its function in connecting with storage facilities and distribution centers. This integration is crucial for connecting primary, large-scale transport systems with the end-users reliant on a stable energy supply.

Nonetheless, road and rail networks present intrinsic challenges. Their restricted capacity relative to pipelines and maritime vessels frequently leads to elevated per unit transportation costs, potentially affecting the overall efficacy of energy distribution. Furthermore, the conveyance of flammable or hazardous energy carriers via road and rail poses considerable safety risks. These challenges require stringent regulatory supervision and the utilization of specialized apparatus to mitigate risks and guarantee safe operations. Notwithstanding these challenges, road and rail transportation are essential for the energy sector, offering the flexibility and localized connectivity crucial for a robust and resilient energy infrastructure.

1.4.5 Emerging and Alternative Systems

As the energy landscape transforms, innovative technologies and methodologies for energy transport are arising to improve efficiency, minimize environmental impacts, and integrate effortlessly with renewable energy sources. Hydrogen and synthetic fuel logistics represent an auspicious sector. Hydrogen is increasingly acknowledged as a crucial energy carrier in the shift towards a low-carbon economy, possessing substantial potential to function as a fuel for transportation, industry, and power generation. This commitment has stimulated comprehensive investigation into its production, storage, and transportation [20]. Synthetic fuels, created by merging hydrogen with captured carbon dioxide, provide a carbon–neutral alternative to traditional fossil fuels, offering a feasible method for diminishing greenhouse gas emissions. Hydrogen can be transported in either compressed or liquefied states. Both methods necessitate specialized infrastructure to effectively mitigate the risks linked to low temperatures and the high reactivity of hydrogen. Alternatively, hydrogen may be transformed into chemical carriers, such as ammonia or methanol, which subsequently utilize existing shipping and pipeline infrastructures. This method incurs extra energy conversion losses during reconversion at the point of use; however, it leverages the established transport infrastructure, enabling a more seamless transition to hydrogen-based energy systems [28].

Concurrently, decentralized energy transport and storage systems are gaining prominence as the emergence of distributed generation, especially from renewable sources, transforms the conventional energy grid. Microgrids, localized battery storage systems, and vehicle-to-grid (V2G) technologies are transforming the storage, management, and transportation of energy in urban and rural settings. These decentralized systems provide substantial benefits regarding grid flexibility by diminishing reliance on centralized energy transport networks. They facilitate localized equilibrium of supply and demand, thereby augmenting overall grid resilience and reliability [29]. Furthermore, decentralized energy systems are ideally positioned to incorporate intermittent renewable energy sources, such as solar and wind, thus diminishing dependence on fossil fuels and fostering a more sustainable energy future. These emerging methodologies in hydrogen logistics and decentralized transport address current challenges in energy distribution and facilitate the development of a more adaptive and environmentally sustainable energy infrastructure [30].

1.5 Sustainable Energy Carriers

Sustainable energy carriers are positioned to be crucial in the worldwide shift towards a low-carbon energy system. The carriers encompassing electricity, hydrogen, ammonia, and diverse biofuels facilitate the storage, transportation, and utilization of energy obtained from renewable sources. Through their role as intermediaries between energy production and end-use, sustainable energy carriers enable the reduction of carbon emissions in several sectors such as electricity generation, transportation, heating, and industry. The development and deployment of sustainable energy carriers are increasingly recognized as crucial elements of a comprehensive energy strategy as the world strives to mitigate climate change and decrease reliance on fossil fuels.

1.5.1 Electricity as an Energy Carrier

Electrical power is the predominant energy medium globally, and its significance in the energy transition is pivotal. Electrical energy can be produced from several sustainable sources, such as solar, wind, hydro, and geothermal energy, and can be distributed over extended distances via power networks. The adaptability of electricity as an energy medium enables its utilization in a diverse array of applications, encompassing the provision of power to residential and commercial establishments, the propulsion of electric vehicles, and the electrification of industrial operations [3].

An inherent benefit of electricity as a means of energy transportation is its capacity to be produced, conveyed, and used with little ecological consequences when obtained from renewable sources. Nevertheless, the incorporation of a substantial proportion of renewable energy into the power system poses difficulties because of the sporadic and fluctuating characteristics of sources such as wind and solar power. Efforts are underway to tackle these issues by developing energy storage technologies, including batteries and pumped hydrostorage, which can store surplus electricity during times of high production and release it when demand surpasses supply [28].

Furthermore, apart from its direct application, electricity can also be harnessed to generate other environmentally friendly energy carriers, such as hydrogen, by chemical processes like electrolysis. This mechanism facilitates the transformation of sustainable energy into forms that can be stored for extended durations or utilized in industries where direct electrification is difficult.

1.5.2 *Hydrogen as an Energy Carrier*

Hydrogen is rapidly becoming recognized as a highly promising sustainable energy carrier, capable of reducing carbon emissions in sectors that are challenging to electrify, including heavy industry, aviation, and long-distance transportation. The production of hydrogen can be achieved from several sources, such as renewable electricity by water electrolysis, natural gas by steam methane reforming with carbon capture and storage (CCS), and biomass gasification [20]. Hydrogen, when derived from renewable sources, is commonly known as green hydrogen, offering the benefit of being a fully emission-free fuel.

The adaptability of hydrogen as an energy carrier stems from its capacity to exist in several states, including pure gas, liquid, and even as a constituent of chemical compounds such as ammonia. It can be stored in substantial amounts, conveyed over extended distances by pipelines or tankers, and transformed into electricity or heat by fuel cells or combustion reactions. Using hydrogen as a raw material for synthesizing synthetic fuels has the potential to supplant fossil fuels in current infrastructure [28].

Nevertheless, the extensive implementation of hydrogen as an energy carrier encounters various obstacles. Key challenges include substantial investments in production infrastructure, such as electrolysis facilities, and establishing a worldwide hydrogen distribution network. Furthermore, the storage and transportation of hydrogen necessitates technological progress to tackle concerns pertaining to effectiveness, safety, and cost.

1.5.3 *Ammonia as an Energy Carrier*

Ammonia, generated by the combination of hydrogen with nitrogen derived from the atmosphere, is increasingly being recognized as a promising sustainable energy carrier, especially for maritime transportation and industrial uses. Ammonia possesses a greater energy density compared to hydrogen, hence facilitating its storage and transportation. Hydrogen can be utilized either as a fuel in internal combustion engines or as a medium for transporting hydrogen, which can be liberated at the location of use by a process known as cracking [29].

An inherent advantage of ammonia is its preexisting infrastructure, as it is already extensively manufactured and transported for application in fertilizers and chemicals. This existing infrastructure has the potential to be modified in order to facilitate the energy transition, so minimizing the requirement for completely new systems. Furthermore, the process of ammonia combustion does not generate carbon dioxide, so establishing it as a zero-carbon fuel upon its production from renewable hydrogen.

Nevertheless, the utilization of ammonia as an energy carrier also presents certain difficulties. A hazardous and corrosive compound, ammonia necessitates meticulous management and storage. The advancement of effective and economical techniques for the production, storage, and use of ammonia will be crucial for its function as a sustainable energy carrier [31].

1.5.4 Biofuels as Energy Carriers

A further category of sustainable energy carriers is biofuels, which are produced from organic materials such as plant biomass, agricultural residues, and waste. Their application lies in the production of liquid fuels, such as ethanol and biodiesel, which have the potential to substitute gasoline and diesel in transportation directly. The compatibility of biofuels with current internal combustion engines and fueling infrastructure renders them a viable short-term remedy for mitigating carbon emissions in the transportation industry [32].

Advanced biofuels, such as cellulosic ethanol and biodiesel derived from algae, provide enhanced sustainability prospects by utilizing non-food raw materials and generating reduced life cycle emissions. Furthermore, biofuels have the capability to generate biogas, which can then be transformed into biomethane and either injected into natural gas pipelines or utilized for the purpose of generating electricity [33].

Biofuel production, however, must be meticulously controlled to prevent adverse environmental and social consequences, including deforestation, biodiversity loss, and competition with food production. Implementing sustainable land use practices and advancing the production of next-generation biofuels that are not dependent on food crops are crucial for ensuring that biofuels make a beneficial contribution to the energy transition [34].

1.5.5 Synthetic Fuels and Other Energy Carriers

The production of synthetic fuels, commonly referred to as electrofuels or e-fuels, involves the combination of hydrogen with carbon dioxide to generate hydrocarbons capable of substituting traditional fossil fuels in the domains of transportation and industry. Renewable electricity can be used to synthesize these fuels, so rendering them a potentially carbon–neutral alternative. The utilization of synthetic fuels in current engines and infrastructure offers a viable approach to reducing carbon emissions in industries such as aviation and shipping, where direct conversion to electric power is difficult [35].

Apart from hydrogen, ammonia, and biofuels, other promising energy carriers encompass methanol, derived from biomass or carbon dioxide and hydrogen, as well as metal fuels, which store energy as metallic powders that can be oxidized to liberate energy. Each of these carriers presents distinct benefits and obstacles, and their progress will rely on advancements in technology, infrastructure, and policy subsidies.

The development and implementation of sustainable energy carriers are crucial for attaining a profound reduction of carbon emissions throughout the worldwide energy system. Nevertheless, a number of obstacles persist. These factors encompass the exorbitant expenses associated with manufacturing and infrastructure, the necessity for technological progress in storage and transportation, and the implementation of regulatory frameworks and quality standards to guarantee safety and effectiveness [36].

Research and innovation will be crucial in surmounting these obstacles. Technological advancements in electrolysis have the potential to decrease the expenses associated with green hydrogen production greatly. Similarly, enhancements in carbon capture and utilization could facilitate the large-scale manufacturing of synthetic fuels. In addition, it will be necessary to enhance international collaboration and investment in order to establish the worldwide infrastructure necessary to facilitate the extensive use of sustainable energy carriers. Ultimately, sustainable energy carriers play a pivotal role in the shift towards a low-carbon energy system. Their capacity to store, transport, and distribute energy derived from renewable sources will be crucial for reducing carbon emissions in several sectors of the economy. This book will undertake a more comprehensive examination of the function of these carriers within the wider framework of energy storage and transportation, emphasizing their capacity to facilitate a sustainable and robust energy future.

1.6 Boil-Off Gas Phenomena

Boil-off gas (BOG) phenomena pose a significant obstacle in the storage, transportation, and processing of liquefied energy carriers and fuels. Given the growing dependence of the global energy system on liquefied energy sources like LNG, liquefied hydrogen (LH_2), liquefied petroleum gas (LPG), and liquefied biofuels, it is crucial to comprehend and control the balance of BOG in order to guarantee the effectiveness, safety, and long-term viability of these energy carriers.

In the process of partial vaporization of the liquid fuel, liquefied gases stored at cryogenic temperatures absorb ambient heat, resulting in the production of boil-off gas. This phenomenon arises due to the inability of even the most well-insulated storage tanks and transport vessels to fully prevent the gradual infiltration of heat. At elevated temperatures, a portion of the liquid fuel undergoes evaporation, forming vapor or gas inside the storage container. The gas, called boil-off gas, raises the pressure within the container, necessitating careful control to avoid safety risks or spoilage of the stored content [37].

Fuels stored at extremely low temperatures, such as LNG (typically stored at around − 162 °C) and liquid hydrogen (stored at an even lower temperature of around − 253 °C), are especially relevant for producing BOG. Because of the critical cryogenic conditions needed to keep these fuels in liquid state, the rate at which they boil-off can be substantial, particularly during extended periods of storage or transportation. When not effectively controlled, BOG can result in financial losses, inefficiencies in operations, and higher levels of greenhouse gas emissions [38].

1.6.1 Boil-Off Gas in LNG

Increasingly important in global energy commerce, LNG is one of the most extensively used liquefied energy carriers. LNG is stored and transported in specialized tanks and vessels specifically engineered to preserve its extremely low temperature. Nevertheless, when LNG undergoes heat absorption during storage or transportation, it produces BOG, which needs to be resolved in order to ensure the safety and economic feasibility of LNG operations.

The management of BOG in LNG shipping can be approached through various methods. An established technique is to use the BOG as fuel for the ship's engines, which has become a customary practice in contemporary LNG carriers fitted with dual-fuel engines. Implementing this method not only reduces the hazards linked to pressure accumulation but also offers a cost-efficient and ecologically sustainable means of propelling the ship [39]. Conversely, BOG can be reliquefied by onboard or shore-based reliquefication systems, but this procedure necessitates substantial energy consumption and can incur high costs.

BOG in land-based LNG storage facilities is typically reliquefied and returned to the storage tank, or compressed and supplied to the natural gas grid. Insufficient capture and utilization of the BOG may result in its venting or flare, so generating methane emissions, which possess a far greater global warming potential compared to CO_2. Thus, it is essential to efficiently manage BOG in order to minimize the environmental consequences of LNG operations [40].

1.6.2 Boil-Off Gas in Liquid Hydrogen

LH_2 is becoming recognized as a crucial energy carrier in the shift towards a low-carbon energy system, namely for use in transportation, industry, and electricity production. Nevertheless, the handling of BOG in LH_2 poses considerable difficulties partly because of the very low boiling point and high diffusivity of hydrogen.

The low boiling point of hydrogen results in significant BOG production even with minimal heat ingress. Furthermore, the low density of hydrogen necessitates its storage in substantial quantities, so augmenting the likelihood of heat absorption and boil-off. The effective control of BOG in LH_2 storage and transport systems necessitates the use of sophisticated insulation methods and active cooling systems to minimize the penetration of heat and decrease the rate at which boiling occurs [41].

Effective management of BOG in LH_2 transportation is especially difficult because of the requirement for prolonged storage and the possibility of prolonged

1.6 Boil-Off Gas Phenomena

transportation durations. Effective management of the BOG is crucial in space missions, particularly when LH_2 is used as rocket fuel, to guarantee fuel availability during launch. The use of LH_2 for large-scale energy applications presents similar difficulties, as boil-off can result in substantial energy losses and safety issues if not adequately resolved [42].

1.6.3 Boil-Off Gas in Other Liquefied Energy Carriers

Apart from LNG and LH_2, other liquefied energy carriers, including LPG and liquefied biofuels, also undergo BOG phenomena. LPG, primarily composed of propane and butane, is stored at temperatures of approximately $-42\,°C$. Although the boil-off rate for LPG is typically lower than that for LNG or LH_2, it is still necessary to control BOG in order to avoid elevated pressure and guarantee safety throughout storage and transportation.

The management of BOG is a significant challenge for liquefied biofuels, such as bio-LNG or bio-DME. Anticipated growth in the production and utilization of these biofuels is a key component of endeavors to reduce carbon emissions in the energy system, namely in industries such as shipping and heavy-duty transportation. Effective management of biofuel storage and transportation will need the adoption of comparable strategies to those employed for fossil-based liquefied fuels while also taking into account the sustainability and life cycle emissions of the biofuels [43].

1.6.4 Strategies for Managing Boil-Off Gas

Efficient and sustainable operation of liquefied energy carriers critically relies on the effective management of BOG. Multiple approaches are used to reduce and optimize BOG in various applications as shown in Fig. 1.3.

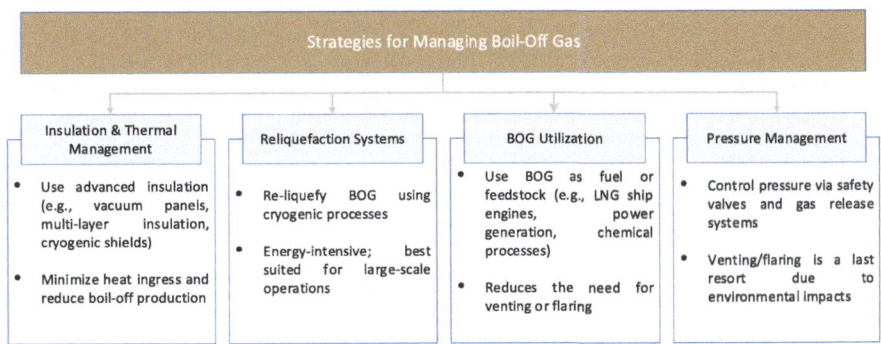

Fig. 1.3 Strategies to manage BOG in various applications

The effective control of BOG continues to be a major obstacle in the utilization of liquefied energy carriers. The increasing demand for LNG, LH_2, and other liquefied fuels necessitates the implementation of efficient, cost-effective, and environmentally sustainable BOG management strategies. Exploration and advancement in insulation technologies, cryogenic systems, and BOG utilization methods will be crucial for surmounting these obstacles and guaranteeing the secure and effective use of liquefied energy carriers.

Ultimately, the boil-off gas phenomenon is a crucial factor to take into account when thinking about the storage and transportation of liquefied energy carriers. Robust management of BOG is crucial to guarantee the safety, efficiency, and sustainability of these fuels in the context of the worldwide energy transition. This book will comprehensively examine the consequences of BOG phenomena for various energy carriers and emphasize methods for mitigating its influence on the energy system.

References

1. IPCC (2023) Climate change 2021—the physical science basis. https://doi.org/10.1017/9781009157896
2. UNFCCC (2015) The Paris agreement
3. IEA (2023) World energy outlook 2023. Paris
4. IRENA (2023) Renewable energy statistics 2023. Abu Dhabi
5. Griscom BW, Adams J, Ellis PW, Houghton RA, Lomax G, Miteva DA et al (2017) Natural climate solutions. Proc Natl Acad Sci USA 114:11645–11650. https://doi.org/10.1073/pnas.1710465114
6. UNEP (2023) Emissions gap report 2023
7. IRENA (2020) Renewable power generation costs in 2020. Abu Dhabi
8. GWEC (2023) Global wind report 2023
9. Lyon K (2020) Using the hydropower sustainability tools in world bank group Client Countries. https://doi.org/10.1596/35018
10. IRENA (2017) Geothermal power: technology brief
11. FAO (2023) The State of food and agriculture 2023. Revealing the true cost of food to transform agrifood systems. Rome: FAO. https://doi.org/10.4060/cc7724en
12. IRENA (2014) Wave energy: technology brief
13. EC (2019) What is the European green deal?
14. Zakeri B, Syri S (2015) Electrical energy storage systems: a comparative life cycle cost analysis. Renew Sustain Energy Rev 42:569–596. https://doi.org/10.1016/j.rser.2014.10.011
15. Luo X, Wang J, Dooner M, Clarke J (2015) Overview of current development in electrical energy storage technologies and the application potential in power system operation. Appl Energy 137:511–536. https://doi.org/10.1016/j.apenergy.2014.09.081
16. Jacob R, Hoffmann M, Weinand JM, Linßen J, Stolten D, Müller M (2023) The future role of thermal energy storage in 100% renewable electricity systems. Renew Sustain Energy Transit 4. https://doi.org/10.1016/j.rset.2023.100059
17. Catsaros O (2023) Lithium-ion battery pack prices hit record low of $139/kWh. BloombergNEF
18. Zheng M (2023) The environmental impacts of lithium and cobalt mining. EarthOrg
19. Yang Y, Okonkwo EG, Huang G, Xu S, Sun W, He Y (2021) On the sustainability of lithium ion battery industry—a review and perspective. Energy Storage Mater 36:186–212. https://doi.org/10.1016/j.ensm.2020.12.019

20. Staffell I, Scamman D, Velazquez Abad A, Balcombe P, Dodds PE, Ekins P et al (2019) The role of hydrogen and fuel cells in the global energy system. Energy Environ Sci 12:463–491. https://doi.org/10.1039/c8ee01157e
21. Khalid M (2024) Smart grids and renewable energy systems: Perspectives and grid integration challenges. Energy Strateg Rev 51. https://doi.org/10.1016/j.esr.2024.101299
22. Semeraro MA (2021) Renewable energy transport via hydrogen pipelines and HVDC transmission lines. Energy Strat Rev 35. https://doi.org/10.1016/j.esr.2021.100658
23. Lu H, Xi D, Qin G (2023) Environmental risk of oil pipeline accidents. Sci Total Environ 874. https://doi.org/10.1016/j.scitotenv.2023.162386
24. Mahajan D, Tan K, Venkatesh T, Kileti P, Clayton CR (2022) Hydrogen blending in gas pipeline networks—a review. Energies 15. https://doi.org/10.3390/en15103582
25. IMO (2023) IMO strategy on reduction of GHG shipping emissions from ships
26. Yarlagadda B, Iyer G, Binsted M, Patel P, Wise M, McLeod J (2024) The future evolution of global natural gas trade. iScience 27. https://doi.org/10.1016/j.isci.2024.108902
27. Sürer MG, Arat HT (2022) Advancements and current technologies on hydrogen fuel cell applications for marine vehicles. Int J Hydrog Energy 47:19865–19875. https://doi.org/10.1016/j.ijhydene.2021.12.251
28. IRENA (2019) Hydrogen: a renewable energy perspective. Abu Dhabi
29. Giddey S, Badwal SPS, Munnings C, Dolan M (2017) Ammonia as a renewable energy transportation media. ACS Sustain Chem Eng 5:10231–10239. https://doi.org/10.1021/acssuschemeng.7b02219
30. Silvast A, Abram S, Copeland C (2023) Energy systems integration as research practice. Technol Anal Strat Manage 35:302–313. https://doi.org/10.1080/09537325.2021.1974376
31. Valera-Medina A, Amer-Hatem F, Azad AK, Dedoussi IC, De Joannon M, Fernandes RX et al (2021) Review on ammonia as a potential fuel: from synthesis to economics. Energy Fuels 35:6964–7029. https://doi.org/10.1021/acs.energyfuels.0c03685
32. Basu P (2013) Production of synthetic fuels and chemicals from biomass. In: Biomass gasification, pyrolysis and torrefaction. Elsevier, pp 375–404. https://doi.org/10.1016/B978-0-12-396488-5.00011-3
33. Skeer J, Boshell F, Ayuso M (2016) Technology innovation outlook for advanced liquid biofuels in transport. ACS Energy Lett 1:724–725. https://doi.org/10.1021/acsenergylett.6b00290
34. Tilman D, Socolow R, Foley JA, Hill J, Larson E, Lynd L et al (2009) Beneficial biofuels—the food, energy, and environment trilemma. Science 325:270–271. https://doi.org/10.1126/science.1177970
35. Schmidt P, Batteiger V, Roth A, Weindorf W, Raksha T (2018) Power-to-liquids as renewable fuel option for aviation: a review. Chem Ing Tech 90:127–140. https://doi.org/10.1002/cite.201700129
36. Blanco H, Faaij A (2018) A review at the role of storage in energy systems with a focus on Power to Gas and long-term storage. Renew Sustain Energy Rev 81:1049–1086. https://doi.org/10.1016/j.rser.2017.07.062
37. Widodo A, Muharam Y (2023) Simulation of boil-off gas recovery and fuel gas optimization for increasing liquefied natural gas production. Energy Rep 10:4503–4515. https://doi.org/10.1016/j.egyr.2023.11.003
38. Kalikatzarakis M, Theotokatos G, Coraddu A, Sayan P, Wong SY (2022) Model based analysis of the boil-off gas management and control for LNG fuelled vessels. Energy 251. https://doi.org/10.1016/j.energy.2022.123872
39. Al-Sobhi SA, AlNouss A, Shamlooh M, Al-Nuaimi K, AlMulla A, Khraisheh M (2021) Sustainable boil-off gas utilization in liquefied natural gas production: economic and environmental benefits. J Clean Prod 296. https://doi.org/10.1016/j.jclepro.2021.126563
40. Shin Y, Lee YP (2009) Design of a boil-off natural gas reliquefaction control system for LNG carriers. Appl Energy 86:37–44. https://doi.org/10.1016/j.apenergy.2008.03.019
41. Hren R, Vujanović A, Van Fan Y, Klemeš JJ, Krajnc D, Čuček L (2023) Hydrogen production, storage and transport for renewable energy and chemicals: an environmental footprint assessment. Renew Sustain Energy Rev 173. https://doi.org/10.1016/j.rser.2022.113113

42. Zhang T, Uratani J, Huang Y, Xu L, Griffiths S, Ding Y (2023) Hydrogen liquefaction and storage: recent progress and perspectives. Renew Sustain Energy Rev 176. https://doi.org/10.1016/j.rser.2023.113204
43. Klass DL (1998) Energy consumption, reserves, depletion, and environmental issues. Biomass Renew Energy Fuels Chem (Elsevier; 1998, pp 1–27). https://doi.org/10.1016/B978-012410950-6/50003-9

Open Access This chapter is licensed under the terms of the Creative Commons Attribution 4.0 International License (http://creativecommons.org/licenses/by/4.0/), which permits use, sharing, adaptation, distribution and reproduction in any medium or format, as long as you give appropriate credit to the original author(s) and the source, provide a link to the Creative Commons license and indicate if changes were made.

The images or other third party material in this chapter are included in the chapter's Creative Commons license, unless indicated otherwise in a credit line to the material. If material is not included in the chapter's Creative Commons license and your intended use is not permitted by statutory regulation or exceeds the permitted use, you will need to obtain permission directly from the copyright holder.

Chapter 2
Natural Gas

2.1 Liquefied Natural Gas

LNG is essential for addressing regional supply–demand discrepancies in a rapidly globalizing energy market. The conversion of natural gas to a liquid state improves its transportability and creates access to markets that lack pipeline infrastructure. The capacity of LNG to offer a more environmentally friendly substitute for other fossil fuels has accelerated its global adoption, in accordance with international efforts to mitigate greenhouse gas emissions. This section offers a thorough academic analysis of LNG, encompassing the technical complexities of liquefaction, the engineering obstacles of cryogenic storage, and global trade dynamics. LNG provides a range of benefits that surpass its ability to reduce volume: (i) LNG enables the cost-effective transportation of natural gas across continents, linking isolated supply sources with distant markets. (ii) By facilitating access to global natural gas reserves, LNG diminishes reliance on regional pipeline infrastructures. (iii) The combustion of natural gas results in reduced emissions of pollutants (e.g., SO_x, NO_x, particulates) and carbon dioxide relative to coal and oil. (iv) LNG is employed in various sectors, such as power generation, industrial manufacturing, and maritime transport, and it facilitates peak-shaving operations and emergency power supplies.

2.1.1 LNG Production

The production of LNG involves a series of complex processes that transform raw natural gas into a storable and transportable liquid, requiring precise regulation of thermodynamic conditions and energy transfers. The process can be divided into three main phases: extraction of natural gas, processing and purification, and liquefaction. Each phase employs sophisticated engineering principles and thermodynamic equations to ensure that the final product complies with rigorous quality and safety standards while optimizing energy efficiency.

The production cycle begins with the extraction of natural gas from underground reservoirs using conventional and unconventional methods. In conventional drilling, vertical wells penetrate porous rock formations, allowing natural gas to flow into the wellbore due to inherent pressure differentials. Nontraditional methods, such as hydraulic fracturing (fracking) and horizontal drilling, are employed to extract gas reserves located within shale or tight formations. The extracted gas is contaminated; it consists of a complex amalgamation of methane (CH_4), water vapor, carbon dioxide (CO_2), hydrogen sulfide (H_2S), and heavier hydrocarbons. The unrefined gas is frequently depicted by the ideal gas law:

$$PV = nRT$$

where P is the pressure, V is the volume, n is the number of moles, R is the universal gas constant, and T is the temperature. This equation serves as a basis for understanding the behavior of the gas under various conditions encountered during extraction and subsequent processing.

Prior to liquefaction, natural gas must undergo processing to eliminate impurities that could harm liquefaction equipment or compromise the quality of the final LNG product. The purification phase includes several essential subprocesses. Initially, dehydration is conducted to eliminate water vapor through glycol absorption or molecular sieves; this is crucial to avert hydrate formation during the cooling process. Subsequently, acid gas removal is executed through amine scrubbing or membrane separation to eradicate CO_2 and H_2S, which are corrosive and can diminish the quality of LNG. Hydrocarbon dew point control is implemented to modify the gas composition and prevent the condensation of heavier hydrocarbons during cooling, while filtration eliminates particulates that may damage downstream equipment.

Liquefaction is the fundamental transformation process in LNG production, wherein purified natural gas is cooled to cryogenic temperatures (approximately −162 °C) and converted into a liquid state. This procedure is energy-demanding and depends on sophisticated refrigeration cycles. A prevalent technique is the cascade cycle, which utilizes several refrigeration stages, each employing distinct refrigerants optimized for particular temperature ranges. A prevalent technique is the mixed refrigerant cycle, which employs a meticulously chosen combination of refrigerants to generate an optimized cooling curve. This method's efficiency is improved as the refrigerant mixture aligns closely with the cooling demands of the natural gas, thereby minimizing exergy losses linked to temperature profile discrepancies. The overall energy efficiency of the liquefaction process is typically represented as the ratio of the minimum theoretical work (derived from reversible processes) to the actual work input.

2.1 Liquefied Natural Gas   27

A third method, the expander (or turbo-expansion) cycle, utilizes the Joule–Thomson effect, wherein a real gas undergoes a temperature variation during expansion at constant enthalpy. During this cycle, the gas undergoes expansion via a turbo-expander, resulting in substantial cooling and the conversion of a portion of its internal energy into mechanical work, thereby enhancing overall process efficiency.

During the liquefaction phase, meticulous regulation of pressure, temperature, and flow rates is achieved by incorporating sophisticated sensors and automated control systems. These systems consistently modify the operational parameters to guarantee that the gas adheres to the intended thermodynamic trajectory, thus reducing energy losses and optimizing LNG production. The stringent application of thermodynamic principles and equations guarantees that LNG production is energy-efficient and adheres to the stringent quality standards necessary for safe storage and transportation.

LNG production is a complex process that integrates both conventional and unconventional extraction methods, stringent processing and purification to eliminate impurities, and sophisticated liquefaction technologies based on essential thermodynamic principles and equations. The production of LNG is fundamentally supported by advanced scientific and engineering principles, encompassing the application of the ideal gas law and Clausius–Clapeyron equation in extraction and purification, as well as the utilization of polytropic process equations and Joule–Thomson coefficients during liquefaction. This comprehensive strategy not only enhances energy efficiency but also guarantees the generation of superior LNG that can be securely stored and transported to satisfy global energy requirements.

2.1.2 LNG Storage

The storage of LNG is a specialized process that utilizes advanced cryogenic systems designed to sustain extremely low temperatures, thereby ensuring that natural gas remains in its liquid form. Central to these systems are cryogenic tanks, which are precisely engineered to reduce heat ingress and thereby regulate the rate of LNG boil-off. The inner containment vessels of these tanks are generally fabricated from premium stainless steel or aluminum alloys. These materials are selected not only for their strength but also for their remarkable capacity to endure the extreme thermal stresses associated with cryogenic temperatures. The design of these inner vessels is paramount, as they must securely contain the LNG while maintaining structural integrity, even under the cyclical thermal expansion and contraction experienced during filling and unloading operations.

An intricate insulation system encircles the inner containment, which is essential for maintaining cryogenic conditions. Multilayer insulation (MLI) and vacuum-insulated panels (VIPs) are frequently utilized to establish a robust barrier against thermal transfer. These insulation systems are designed to minimize thermal conductivity, ensuring that minimal heat enters the storage vessel. This is essential for regulating boil-off gas (BOG), the unavoidable vaporization of LNG resulting from residual heat ingress. By effectively regulating the boil-off rate, these insulation systems preserve LNG at its necessary low temperature, minimizing energy losses and the requirement for ongoing reliquefaction.

The external frameworks of LNG storage tanks offer supplementary layers of safeguarding and utility. These structures are engineered to uphold the inner containment and insulation layers, providing mechanical stability and secondary containment to avert leaks. They are frequently built with materials and design elements that reduce the likelihood of thermal bridges, which can facilitate undesirable heat transfer. Often, these external layers comprise redundant safety barriers that guarantee the LNG remains securely isolated, even if the inner containment is breached. The intricate design of these tanks prioritizes thermal efficiency while also ensuring long-term durability and safety, establishing them as a fundamental component of LNG infrastructure.

Rigorous design considerations and safety protocols are essential at every stage of LNG storage system development to guarantee safe storage. Pressure relief mechanisms serve as a fundamental safety feature; safety valves and venting systems are systematically incorporated to regulate pressure surges caused by boil-off. These systems are engineered to engage automatically under specified conditions, thus averting overpressure situations that may result in catastrophic failure. Engineers perform comprehensive thermal stress analyses to consider the varying expansion and contraction of materials in cryogenic environments. This meticulous evaluation aids in the design of tanks capable of withstanding repeated thermal cycles without fatigue or deterioration.

Advanced leak detection systems constitute an essential element of LNG storage safety. A network of sensors is installed throughout the storage facility to continuously monitor temperature, pressure, and any indications of gas leakage. These sensors are frequently connected to automated monitoring systems that deliver real-time data, enabling operators to address any anomalies promptly. This proactive leak detection method improves operational safety and reduces the environmental risks linked to LNG handling. Moreover, adherence to international standards—such as those established by the American Society of Mechanical Engineers (ASME) and the International Electrotechnical Commission (IEC)—guarantees that the design and operation of LNG storage systems achieve optimal safety and reliability. These protocols collectively support the comprehensive safety framework that regulates LNG storage globally.

2.1.3 LNG Transport

The transportation and distribution of LNG are critical elements of its value chain, requiring specialized systems to preserve LNG's cryogenic state from production sites to end-users. Maritime transport is essential in this network, facilitating the movement of LNG over extensive distances using specialized LNG carriers. These vessels are outfitted with advanced cryogenic tanks designed to reduce heat transfer

2.1 Liquefied Natural Gas

during extended maritime journeys. The carriers are designed with highly efficient insulation systems akin to those utilized in onshore storage facilities, guaranteeing that the LNG is maintained at the necessary low temperatures. Moreover, these vessels frequently feature integrated reliquefication units that can recondense any LNG that vaporizes during transit, thus regulating boil-off gas and enhancing overall energy efficiency.

LNG carriers are augmented by advanced onboard systems, encompassing sophisticated pressure management solutions and real-time monitoring apparatus. These systems consistently monitor the state of the LNG, promptly rectifying any deviations from ideal storage conditions. The incorporation of these technologies not only protects the LNG cargo but also improves the vessel's operational adaptability, enabling it to safely traverse diverse climatic conditions and extended journeys. The maritime sector of LNG transportation integrates advanced cryogenic technology with stringent safety protocols, aimed at facilitating global LNG trade.

Land-based solutions, including cryogenic road and rail tankers, are utilized for regional distribution and short-distance transport. These specialized vehicles are engineered to sustain the cryogenic conditions essential for LNG, utilizing advanced insulation materials and pressure regulation systems to control thermal dynamics during transportation. In contrast to maritime transport, which involves considerable distances and prolonged transit durations, land-based transport frequently necessitates swift and effective delivery systems. Upon arrival at its destination via these tankers, LNG is generally routed to regasification terminals. At these facilities, LNG is reverted to its gaseous state via a regulated process employing heat exchangers and vaporization systems. The regasification process is essential as it facilitates the incorporation of LNG into current natural gas pipeline systems, thus connecting distant LNG production sites with regional or national gas distribution networks.

Notwithstanding the advanced technologies utilized in LNG transportation, the supply chain encounters numerous logistical challenges that require ongoing management. A primary challenge is sustaining cryogenic conditions during transportation. Continuous insulation and real-time monitoring are crucial to avert excessive boil-off and reduce energy losses, ensuring that the LNG retains its liquid state until it arrives at its destination. Moreover, the infrastructure necessary for LNG transportation—including specialized vessels, storage tanks, and regasification facilities—requires substantial capital investment. The elevated expenses require synchronized long-term planning and strategic alliances throughout the global LNG sector. Moreover, international trade regulations enforce rigorous standards regarding safety, environmental conservation, and economic practices, affecting market dynamics and pricing. Successfully navigating regulatory landscapes necessitates a thorough comprehension of international standards and a dedication to upholding the highest standards of operational integrity. Ongoing innovation and strategic management are advancing the transportation and distribution networks for LNG, thereby ensuring its status as a secure and viable element of the global energy supply.

2.2 Compressed Natural Gas

2.2.1 CNG Production

The production of CNG is a complex process that converts raw natural gas into a compressed format appropriate for storage, transportation, and utilization. The process commences at the extraction phase, during which natural gas is obtained from subterranean reservoirs. Extraction can occur via conventional drilling, which accesses porous rock formations, or through advanced methods such as hydraulic fracturing and horizontal drilling that retrieve gas trapped in shale or tight formations. The unrefined natural gas obtained from these sources is generally an intricate mixture of methane and various impurities, including water vapor, carbon dioxide, hydrogen sulfide, and heavier hydrocarbons. The gas composition upon exiting the well is crucial, as it determines the necessary processing steps for compression preparation.

Subsequent to extraction, the natural gas undergoes an extensive processing and purification phase. At this stage, impurities that may compromise compressor performance or the quality of the final product are meticulously eliminated. Dehydration is a primary critical step; water vapor is removed from the gas through techniques such as glycol absorption or molecular sieves to avert the formation of hydrates that may obstruct pipelines and equipment. Subsequently, acid gases, including carbon dioxide and hydrogen sulfide, are removed via processes such as amine scrubbing, thereby enhancing the safety and stability of the gas during compression. Supplementary treatments, including hydrocarbon dew point management and particulate filtration, enhance the refinement of natural gas, yielding a clean fuel optimized for high-pressure storage and transportation.

Upon purification, the gas is routed to the compression phase, where it undergoes conversion into CNG. This phase entails compressing natural gas to pressures generally between 200 and 250 bar, significantly diminishing its volume while maintaining its energy content. The compression process is energy-intensive and must be performed with precision to guarantee efficient and safe gas compaction. Contemporary CNG production facilities frequently employ sophisticated reciprocating or rotary compressors, occasionally arranged in multiple stages with intercooling between stages to enhance energy efficiency and preserve gas integrity. The objective is to attain a stable, high-pressure product that can be efficiently stored in specially engineered cylinders or integrated storage systems without jeopardizing safety or performance.

The ultimate output of CNG production, a highly compressed and purified natural gas, is prepared for storage and distribution. CNG storage systems are designed to securely contain the gas at elevated pressures, utilizing resilient materials like high-strength steel or composite substances that ensure durability and corrosion resistance. These storage solutions are essential for maintaining the stability of compressed gas until required for vehicle fueling, power generation, or other industrial uses. The

2.2 Compressed Natural Gas

complete process—from extraction and purification to compression and storage—is engineered to optimize energy efficiency while adhering to rigorous safety standards and environmental regulations, thus guaranteeing that CNG remains a feasible, economical, and eco-friendly alternative to other fossil fuels.

CNG production constitutes an essential component of the energy value chain. The conversion of natural gas into a compact, transportable form facilitates flexible energy distribution, diminishes greenhouse gas emissions in cleaner applications, and serves as a transition between conventional fossil fuels and future renewable energy systems. Improvements in extraction, processing, and compression technologies enhance efficiency and safety, establishing CNG as a crucial element of contemporary global energy strategies.

2.2.2 CNG Storage

The storage of CNG is a specialized procedure focused on preserving the gas at elevated pressures within durable containment systems. In contrast to LNG, which necessitates cryogenic temperatures, CNG storage entails compressing natural gas to pressures generally between 200 and 250 bar. This high-pressure environment is sustained in specially engineered cylinders designed to minimize gas leakage and ensure structural integrity. These cylinders are frequently fabricated from high-strength materials, including steel or advanced composites. Steel cylinders, recognized for their durability and reliability, have traditionally been the standard; however, the advancement of composite cylinders—comprising a thin metallic liner encased in carbon-fiber or fiberglass overwrap—has facilitated substantial weight reductions and enhanced corrosion resistance. The design of these cylinders must consider the intense mechanical stresses experienced during rapid filling and discharge cycles, ensuring that each unit can safely withstand the internal pressures throughout its operational lifespan.

The CNG storage system, in addition to the fundamental container, includes sophisticated safety and monitoring mechanisms intended to mitigate the inherent hazards associated with high-pressure gas storage. Central to these systems are pressure relief valves and venting mechanisms, designed to activate automatically when internal pressure surpasses predetermined limits. These safety devices prevent potentially hazardous overpressure situations, safeguarding both the storage facility and adjacent areas. The design process entails comprehensive computational and experimental analyses to comprehend material behavior under cyclic pressure. Such analyses guide the formulation of resilient construction methodologies and the choice of materials that demonstrate minimal fatigue and wear under high-pressure circumstances. Moreover, contemporary CNG storage facilities frequently incorporate advanced sensor networks that incessantly monitor variables such as pressure, temperature, and gas composition, thus facilitating the prompt identification of any deviations from optimal operating conditions. Compliance with international safety standards and stringent quality control during manufacturing and inspection phases enhances the reliability of these high-pressure storage systems.

The thorough design of CNG storage also considers practical factors such as maintenance, accessibility, and the incorporation of these systems into broader energy infrastructure. Maintenance protocols are implemented to detect and address any wear or deterioration before it jeopardizes safety or performance. Storage facilities are often designed for modularity, enabling scalability to accommodate fluctuating demand while ensuring consistent performance metrics. The arrangement and positioning of high-pressure cylinders are meticulously designed to optimize space efficiency and enable regular inspections. Furthermore, the incorporation of these storage systems into the extensive energy network requires compatibility with fueling stations, industrial applications, and mobile platforms. This comprehensive design approach guarantees that CNG storage adheres to rigorous safety and operational standards while accommodating the practical requirements of contemporary energy distribution.

2.2.3 CNG Transport

The transportation and distribution of compressed natural gas (CNG) are essential components of its supply chain, encompassing the transfer of high-pressure gas from production sites to end-users via pipeline networks and mobile delivery systems. Pipelines function as the fundamental infrastructure for CNG distribution across extensive distances, offering a consistent and efficient method for transporting compressed gas. These pipeline systems are designed to sustain high pressures while reducing energy losses and ensuring operational safety. The materials used for these pipelines, usually high-grade steel, are chosen for their capacity to endure significant internal pressures and resist corrosion over prolonged durations. Routine maintenance and sophisticated monitoring systems, incorporating pressure sensors and real-time data transmission, facilitate the prompt detection of leaks or structural deficiencies, thereby improving the overall reliability of the pipeline infrastructure.

Besides fixed pipeline networks, mobile transportation is essential for delivering CNG to areas lacking pipeline access or where demand is concentrated. Specialized high-pressure tanker trucks, engineered to securely hold CNG in cylinders or integrated high-pressure vessels, are frequently utilized for this function. These vehicles feature sophisticated insulation and pressure regulation systems to guarantee the secure containment of gas during transit, regardless of fluctuating ambient conditions. The design of these mobile units necessitates a careful equilibrium among durability, weight, and safety. Lightweight composite materials are progressively utilized in the construction of transport tanks, facilitating easier handling and enhancing fuel efficiency while adhering to rigorous safety standards. Moreover, the vehicles are equipped with extensive safety mechanisms, including automatic shut-off valves and emergency venting systems, which facilitate the effective management of unforeseen pressure accumulation or impact-related occurrences.

Upon arrival at its destination, whether via pipelines or mobile transport, CNG is generally supplied to refueling stations or industrial facilities for use as fuel in vehicles, power generation, or other applications. High-pressure dispensing systems are employed at refueling stations to transfer CNG into vehicle tanks, with meticulous control of flow rate and pressure to facilitate swift and safe refueling. These stations feature automated monitoring systems that consistently detect any indications of pressure deviation or leakage, and they comply with stringent regulatory standards to guarantee the safety of both operators and the public. The CNG distribution network is engineered for high responsiveness to demand fluctuations, featuring strategically positioned storage facilities and transport hubs that enable rapid resupply and reduce downtime. This versatile approach to the transportation and distribution of CNG optimizes the logistical chain while ensuring the safe and efficient delivery of high-pressure gas from production sites to various end-uses.

The transportation and distribution systems for CNG embody advanced engineering, stringent safety standards, and strategic planning. Continuous innovation and technological refinement address the challenges of sustaining high-pressure conditions during transit, managing physical stresses on storage and transport vessels, and adhering to international safety and environmental regulations. As the global energy landscape transforms, CNG transportation and distribution infrastructure will inevitably progress, propelled by the dual objectives of improving operational efficiency and guaranteeing optimal safety and environmental protection.

2.3 Synthetic Natural Gas

2.3.1 SNG Production

The production of SNG is a complex, multistage process that transforms carbon-rich feedstocks into a methane-rich gas similar to conventional natural gas. The process commences with the gasification of the selected feedstock, which may include coal, biomass, or municipal waste. During the gasification process, the feedstock is exposed to elevated temperatures within a regulated, oxygen-restricted setting. This thermal conversion decomposes the solid material into a combustible gas mixture termed synthesis gas (syngas), predominantly comprising carbon monoxide and hydrogen, along with minor quantities of carbon dioxide and other trace gases. The quality and composition of syngas are essential, as they dictate the efficiency of the following conversion processes.

After syngas production, it undergoes a comprehensive purification process to eliminate impurities that may jeopardize subsequent reactions or harm processing equipment. Contaminants including tar, particulates, sulfur compounds, and other undesirable chemicals are meticulously eliminated through diverse mechanical and chemical treatment techniques. This purification step is essential, as even minute quantities of impurities can inhibit the catalysts employed subsequently, diminishing

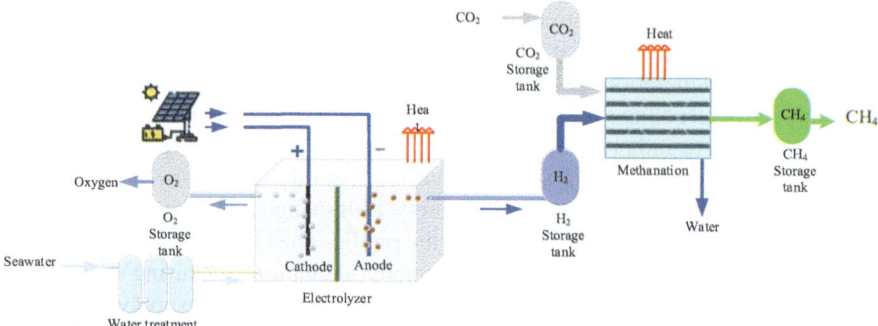

Fig. 2.1 Conversion of renewable energy into SNG (from [1], licensed under CC-BY 4.0)

the efficacy of the methanation reaction. The objective is to achieve a purified gas stream with an ideal composition for methane conversion.

Following purification, the refined syngas is channeled into the methanation reactor, where it is transformed into synthetic natural gas. In this reactor, the hydrogen and carbon monoxide in the syngas react over a catalyst to produce methane and water. In certain configurations, carbon dioxide in the syngas can be transformed into methane through a reaction with supplementary hydrogen. The methanation reaction is exothermic, indicating it releases heat, necessitating meticulous thermal management to sustain optimal reaction conditions. The efficacy of the methanation process is a crucial factor in the overall performance of the SNG production system, as it directly affects the purity and yield of the final product, as shown in Fig. 2.1.

Subsequent to the methanation process, the resultant gas mixture, now augmented with methane, is subjected to additional conditioning and purification to eliminate any remaining water vapor and trace contaminants. This conclusive treatment guarantees that the synthetic natural gas complies with rigorous quality standards, rendering it appropriate for incorporation into current natural gas pipelines and for utilization in diverse end-user applications. The gas is generally cooled, condensed if required, and filtered to yield a stable, high-quality fuel that is nearly indistinguishable from conventional natural gas in performance attributes.

The generation of SNG presents considerable potential for improving energy security and sustainability. A primary advantage of SNG production is its versatility; it can accept a diverse range of feedstocks, including those not conventionally utilized in energy generation. This adaptability not only enhances the energy portfolio but also offers a means to transform low-value or waste materials into a valuable energy resource. Furthermore, synthetic natural gas can play an essential role in power-to-gas systems, wherein excess renewable energy is utilized to generate hydrogen through water electrolysis. This hydrogen can subsequently react with captured carbon dioxide in the methanation process to produce methane, thereby storing renewable energy in a chemical form that is readily transportable and distributable.

Notwithstanding its numerous benefits, SNG production encounters various challenges. The preliminary gasification process demands significant energy, and attaining high conversion efficiency necessitates meticulous regulation of reaction conditions. The process's overall efficiency is determined by the quality of the feedstock, the configuration of the gasification and purification systems, and the efficacy of the methanation catalysts. Economic factors, including the accessibility and expense of feedstocks, capital investment necessities, and market competition with traditional natural gas, significantly influence the viability of large-scale SNG production. Moreover, environmental factors—specifically the handling of by-products and the reduction of greenhouse gas emissions—are essential for ensuring that SNG production positively impacts a sustainable energy future.

In summary, SNG production is a complex process that entails transforming various carbonaceous feedstocks into methane-rich gas via a sequence of stages, including gasification, purification, and catalytic methanation. Synthetic natural gas converts waste materials or unconventional feedstocks into a valuable energy carrier, providing a promising avenue to improve energy security, diminish dependence on fossil fuels, and store renewable energy. Continuous progress in gasification technologies, catalyst innovation, and process integration is crucial for addressing existing challenges and realizing the complete potential of SNG as a sustainable and economically feasible energy resource.

2.3.2 SNG Storage

The storage of SNG is essential to its function as a multifaceted energy carrier. Synthetic methane, chemically similar to conventional natural gas (primarily methane), is generated via the gasification of feedstocks, including coal, biomass, or municipal waste, and is succeeded by methanation processes. Due to its analogous properties, SNG can be stored utilizing technologies similar to those used for natural gas, especially high-pressure containment systems. Storage vessels for SNG are generally engineered to sustain high-pressure conditions, typically between 200 and 250 bar, to attain elevated energy density and facilitate efficient delivery to end-users. These storage systems are fabricated from high-strength materials, including specialized steels or advanced composites. In numerous contemporary installations, a slender metallic liner is fortified with a composite overwrap to enhance durability, decrease weight, and improve corrosion resistance. The design of these vessels must meticulously consider the mechanical stresses experienced during rapid filling and discharge cycles, along with the cyclical pressure fluctuations that arise during extended operational periods.

The storage of SNG necessitates utmost safety and reliability. The design incorporates advanced safety features, including pressure relief valves and venting systems, which are engineered to release gas when internal pressure surpasses established safe thresholds automatically. These safety systems are underpinned by comprehensive engineering analyses that evaluate the dynamic behavior of materials in high-pressure environments. The storage units are outfitted with sophisticated sensor networks that incessantly monitor essential parameters, such as pressure, temperature, and gas composition. This real-time data acquisition enables operators to identify deviations from optimal conditions and promptly implement corrective actions. Moreover, the construction of SNG storage facilities complies with stringent international standards, including those set by the ASME and other regulatory authorities, thereby ensuring the utmost quality and safety are upheld. These extensive measures not only safeguard the stored gas but also establish a resilient framework for sustained operational reliability across diverse industrial and urban environments.

Besides high-pressure cylinder storage, SNG can be incorporated into existing pipeline networks, functioning as a type of distributed storage. Pipelines, typically made from high-quality steel, are engineered to manage the continuous transport of high-pressure gas across long distances. These systems necessitate meticulous thermal and mechanical management to guarantee the stability of the stored SNG despite variations in ambient conditions. Routine inspection and maintenance procedures, coupled with digital monitoring systems, are essential for maintaining the integrity of fixed storage vessels and pipeline systems. The storage infrastructure for SNG embodies a comprehensive strategy that integrates advanced materials, innovative engineering designs, and rigorous safety standards, guaranteeing synthetic natural gas's availability and security throughout its life cycle.

2.3.3 SNG Transport

The transportation and distribution of SNG constitute a vital aspect of its supply chain, guaranteeing that gas generated from various feedstocks arrives at its designated market efficiently and securely. Since SNG is chemically similar to conventional natural gas, it is predominantly transported via the same established networks, particularly through high-pressure pipelines and specialized mobile delivery systems. The long-distance transport of SNG is primarily conducted through a comprehensive pipeline network designed to sustain the high pressures required for effective gas flow. These pipelines are fabricated from high-strength materials and include numerous safety features, such as sophisticated pressure sensors, automated control systems, and routine inspection protocols. The incorporation of SNG into current natural gas pipeline systems bolsters energy security and facilitates market adaptability by connecting distant production sites with urban areas and industrial centers. This seamless integration is facilitated by digital monitoring and control systems that deliver real-time data on gas flow, pressure, and temperature, ensuring that any deviations from safe operating conditions are promptly identified and rectified.

In addition to the fixed pipeline network, mobile transportation solutions are vital in areas with limited pipeline connectivity or where swift delivery is crucial. High-pressure tanker trucks and railcars, engineered for the secure transport of compressed gases, are frequently utilized for the regional distribution of synthetic methane. These vehicles are constructed with durable containment systems similar to those in stationary storage facilities, employing advanced composite materials and high-quality steel to sustain the necessary high-pressure conditions. They possess cutting-edge safety features, such as emergency shut-off valves and automatic pressure relief mechanisms, which assist in risk mitigation during transit. The design of these mobile units meticulously balances durability, ease of handling, and adherence to stringent regulatory standards. Upon reaching distribution centers or end-user facilities, SNG is generally dispensed via specialized high-pressure systems that convey the gas into local pipelines, storage tanks, or directly into end-use apparatus such as power generation units or industrial burners.

The supply chain for SNG is a complex and dynamic system that must address various logistical challenges to guarantee continuous and reliable delivery. A principal challenge is preserving the integrity of high-pressure conditions during the transportation process from the production site to the end consumer. This necessitates the utilization of advanced materials and engineering designs, as well as a coordinated strategy for maintenance and safety supervision. Regular monitoring of pressure, temperature, and flow rates—combined with automated control systems—ensures that the integrity of the SNG is maintained at every stage of its journey. Furthermore, the economic viability of SNG transportation is closely linked to the optimization of infrastructure investments, which include both the expansion of pipeline networks and the acquisition of specialized mobile transport vehicles. Navigating the complex landscape of international safety and environmental regulations is also critical, as these standards influence market dynamics, pricing, and operational practices. Through continuous innovation and strategic planning, the transportation and distribution systems for SNG are evolving to meet the dual imperatives of operational efficiency and environmental sustainability, thereby reinforcing SNG's position as a secure and adaptable energy resource for the future.

Reference

1. Al-Breiki M, Bicer Y (2023) Techno-economic evaluation of a power-to-methane plant: levelized cost of methane, financial performance metrics, and sensitivity analysis. Chem Eng J 471:144725. https://doi.org/10.1016/j.cej.2023.144725

Open Access This chapter is licensed under the terms of the Creative Commons Attribution 4.0 International License (http://creativecommons.org/licenses/by/4.0/), which permits use, sharing, adaptation, distribution and reproduction in any medium or format, as long as you give appropriate credit to the original author(s) and the source, provide a link to the Creative Commons license and indicate if changes were made.

The images or other third party material in this chapter are included in the chapter's Creative Commons license, unless indicated otherwise in a credit line to the material. If material is not included in the chapter's Creative Commons license and your intended use is not permitted by statutory regulation or exceeds the permitted use, you will need to obtain permission directly from the copyright holder.

Chapter 3
Hydrogen

3.1 Hydrogen Production

Hydrogen has become an essential element in the global energy transition, significantly contributing to carbon neutrality and the diversification of energy supply. Hydrogen, as a clean and efficient energy carrier, can be generated from various feedstocks through diverse technologies, facilitating its utilization across numerous sectors, including transportation, power generation, and industrial manufacturing.

The importance of hydrogen resides in its capacity to substitute fossil fuels in energy-intensive industries while mitigating greenhouse gas emissions. Hydrogen can be classified into various types, such as fossil, renewable, and nuclear, based on the production method and energy source. Each of these pathways possesses unique technological, environmental, and economic ramifications.

Figure 3.1 describes the principal hydrogen production pathways, categorized according to their feedstocks and processing methods. It emphasizes the principal techniques including steam methane reforming (SMR), coal gasification, electrolysis, biomass reforming, and thermochemical cycles. The diagram differentiates among fossil hydrogen, fossil hydrogen with diminished emissions (attained via carbon capture and storage), renewable hydrogen produced through electrolysis powered by solar and wind energy, and nuclear hydrogen. Furthermore, the image underscores the significance of hydrogen liquefaction (LH_2), which is essential for storage and transportation.

Comprehending these production pathways is crucial for evaluating the viability of hydrogen as a large-scale energy solution. This chapter examines the technological principles underlying each hydrogen production method, its efficiency, environmental implications, and future prospects within the global energy framework. By analyzing these factors, we can more effectively assess hydrogen's contribution to a sustainable and decarbonized future.

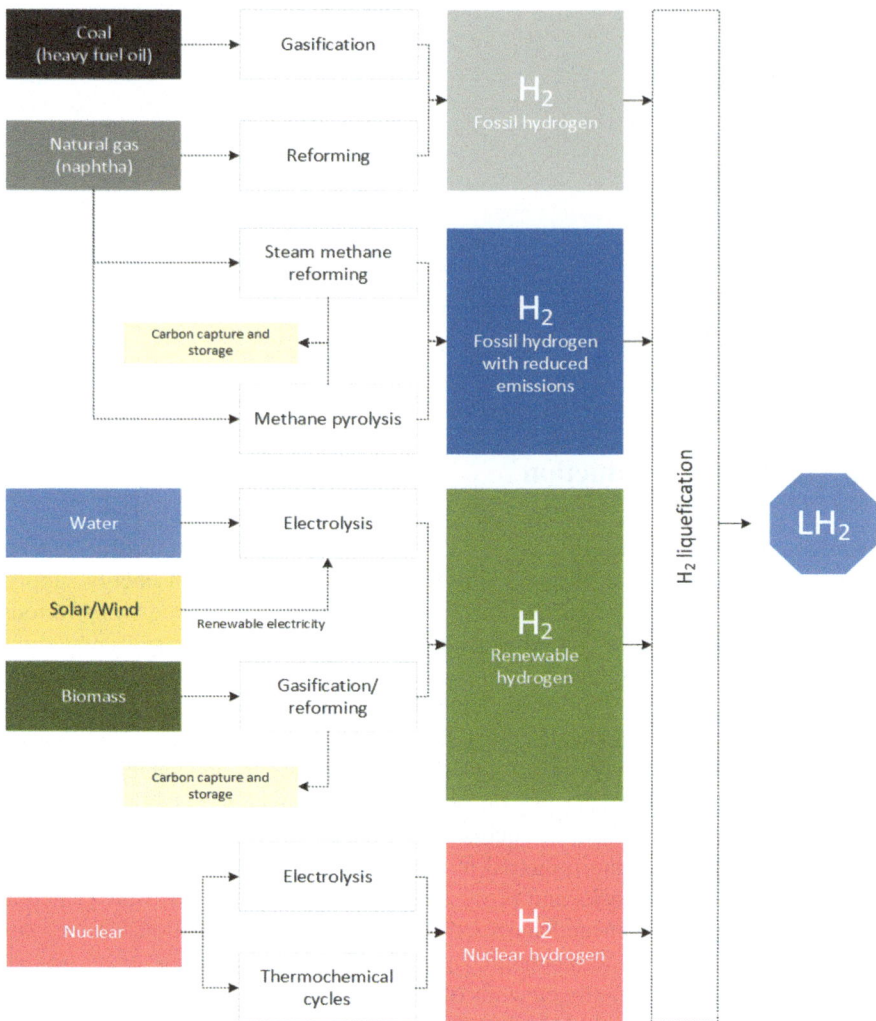

Fig. 3.1 Overview of hydrogen production pathways and classification

3.1.1 Steam Methane Reforming

The production of hydrogen through steam methane reforming (SMR) is a multi-step process that begins with the desulfurization of natural gas to remove sulfur compounds, which could poison the catalysts used in later stages. In this step, natural gas, primarily methane (CH_4), is treated with hydrogen (H_2) to convert sulfur compounds into hydrogen sulfide (H_2S), yielding desulfurized natural gas.

Following desulfurization, the core process of steam reforming takes place. Here, the desulfurized methane reacts with steam (H_2O) at high temperatures (700–1000

3.1 Hydrogen Production

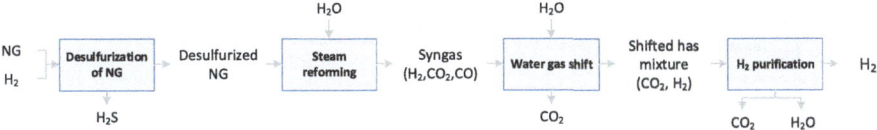

Fig. 3.2 Steps to produce hydrogen using steam methane reforming

°C) in the presence of a nickel-based catalyst. This endothermic reaction primarily produces a mixture of hydrogen (H_2) and carbon monoxide (CO), according to the equation:

$$CH_4 + H_2O \rightarrow CO + 3H_2.$$

Additionally, a secondary reaction occurs, forming carbon dioxide (CO_2) and more hydrogen:

$$CH_4 + 2H_2O \rightarrow CO_2 + 4H_2.$$

The resulting gas mixture, known as syngas, then undergoes the water–gas shift reaction to maximize hydrogen yield. In this step, carbon monoxide reacts with steam to produce additional hydrogen and carbon dioxide in an exothermic reaction:

$$CO + H_2O \rightarrow CO_2 + H_2.$$

This reaction typically occurs in two stages, at high and low temperatures, to optimize efficiency.

Finally, the hydrogen is purified from the shifted gas mixture using techniques such as pressure swing adsorption (PSA). This step separates hydrogen from other components like carbon dioxide and residual carbon monoxide, resulting in high-purity hydrogen suitable for various applications. Throughout the SMR process, the key chemical reactions involve the transformation of methane and water into hydrogen, carbon monoxide, and carbon dioxide, facilitated by specific catalysts and controlled conditions. Figure 3.2 illustrates steps to produce hydrogen using steam methane reforming.

3.1.2 Methane Pyrolysis

In methane pyrolysis in other words methane decomposition or methane cracking, natural gas, primarily composed of methane (CH_4), is separated by application of high-grade heat. Natural gas is initially prepared by ensuring it is free from impurities. The core process involves heating the methane to high temperatures (above 1000 °C) in the absence of oxygen, causing it to decompose into hydrogen gas (H_2) and solid

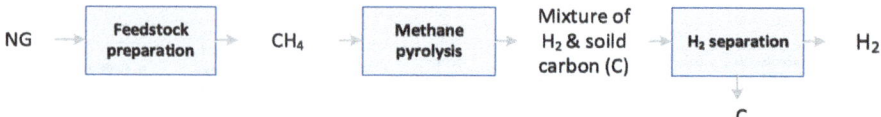

Fig. 3.3 Steps to produce hydrogen using methane pyrolysis

carbon (C). This endothermic reaction is represented by the equation:

$$CH_4 \rightarrow C + 2H_2$$

The resulting mixture of hydrogen gas and solid carbon is then separated, typically through filtration or other gas–solid separation methods, to yield pure hydrogen gas. The solid carbon is collected separately for storage or industrial applications. Figure 3.3 illustrates steps to produce hydrogen using methane pyrolysis. Methane pyrolysis thus efficiently produces hydrogen while simultaneously generating solid carbon, which can have various uses.

3.1.3 Partial Oxidation of Hydrocarbons

The production of hydrogen from hydrocarbons (e.g., oil) through partial oxidation involves several key steps. First, the hydrocarbon feedstock (oil) undergoes pretreatment to remove impurities such as sulfur compounds, which could poison the catalysts used in later stages. The core process, partial oxidation, takes place at high temperatures (typically 1200–1500 °C) where the pretreated hydrocarbon feedstock reacts with a controlled amount of oxygen. This exothermic reaction produces syngas, a mixture of hydrogen (H_2), carbon monoxide (CO), and carbon dioxide (CO_2). The exact stoichiometry depends on the specific hydrocarbon, but a general reaction for a hydrocarbon, C_nH_m, is:

$$C_nH_m + \frac{n}{2}O_2 \rightarrow nCO + \frac{m}{2}H_2.$$

An additional reaction involving steam (H_2O) can also occur:

$$C_nH_m + \frac{n}{2}O_2 + H_2O \rightarrow nCO_2 + \frac{m}{2}H_2.$$

Following this, the syngas undergoes the water–gas shift reaction where carbon monoxide reacts with steam at high temperatures (HTS at about 350 °C and LTS at about 200 °C) to produce additional hydrogen and carbon dioxide, represented by

$$CO + H_2O \rightarrow CO_2 + H_2.$$

3.1 Hydrogen Production

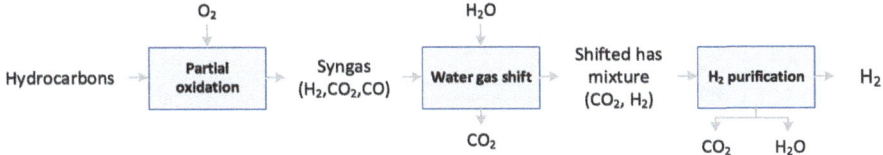

Fig. 3.4 Steps to produce hydrogen from hydrocarbons using partial oxidation

Finally, the resulting gas mixture is purified using techniques such as pressure swing adsorption (PSA) to separate hydrogen from other gases, yielding high-purity hydrogen. This comprehensive process efficiently converts hydrocarbons into hydrogen while managing the by-products through controlled reactions and separation methods. Figure 3.4 illustrates steps to produce hydrogen from hydrocarbons using partial oxidation.

3.1.4 Plasma Reforming of Hydrocarbons

The production of hydrogen from hydrocarbons using plasma reforming involves several critical steps, each with specific inputs, outputs, and conditions. Initially, the hydrocarbon feedstock (such as natural gas or oil) undergoes pretreatment to remove impurities, such as sulfur compounds, which could interfere with the reforming process. Next, plasma is generated using electrical energy to create a highly ionized gas. Technologies like arc discharge or microwave discharge are used to achieve this, resulting in a plasma that reaches temperatures of several thousand degrees Celsius. This plasma provides the necessary environment for the reforming reaction. The core process, plasma reforming, involves introducing the pretreated hydrocarbon feedstock into the plasma. The extremely high temperatures and energy levels within the plasma cause the hydrocarbons to break down into smaller molecules. For example, methane (CH_4) can decompose into hydrogen (H_2) and solid carbon (C) through the reaction:

$$CH_4 \rightarrow C + 2H_2.$$

Additionally, methane can react with steam (H_2O) in the plasma to produce syngas, a mixture of hydrogen and carbon monoxide (CO):

$$CH_4 \rightarrow H_2O \rightarrow CO + 3H_2.$$

Further reaction with excess steam can yield carbon dioxide (CO_2) and more hydrogen:

$$CH_4 \rightarrow 2H_2O \rightarrow CO_2 + 4H_2.$$

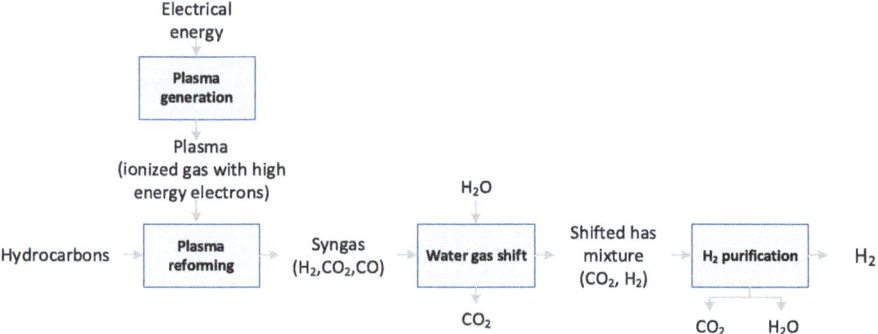

Fig. 3.5 Steps to produce hydrogen from hydrocarbons using plasma reforming

Following the plasma reforming, the syngas produced (comprising hydrogen, carbon monoxide, and carbon dioxide) undergoes the water–gas shift reaction. In this step, carbon monoxide reacts with steam at high temperatures to produce additional hydrogen and carbon dioxide. Finally, the resulting gas mixture is purified. Using techniques such as pressure swing adsorption (PSA), hydrogen is separated from other gases, including carbon dioxide and residual carbon monoxide, resulting in high-purity hydrogen suitable for various applications. This comprehensive process leverages the high-energy environment of plasma to efficiently break down hydrocarbons, producing hydrogen and managing by-products through controlled reactions and separation methods. Figure 3.5 illustrates steps to produce hydrogen from hydrocarbons using plasma reforming.

3.1.5 Coal Gasification

The production of hydrogen from coal using coal gasification involves several key steps, each with specific inputs, outputs, and conditions. Initially, raw coal is prepared by crushing and grinding it into a fine powder, increasing its surface area to enhance the efficiency of subsequent reactions. The core process, gasification, involves reacting the pulverized coal with oxygen (O_2) and steam (H_2O) at high temperatures, typically ranging from 800 to 1800 °C, in a gasifier. This high-temperature environment facilitates the conversion of coal into syngas, a mixture of hydrogen (H_2), carbon monoxide (CO), and carbon dioxide (CO_2). The primary reactions involved in the gasification process are:

Combustion Reaction

$$C + O_2 \rightarrow CO_2.$$

3.1 Hydrogen Production

This exothermic reaction generates heat, which helps maintain the high temperature required for other reactions.

Partial Oxidation Reaction

$$C + \frac{1}{2}O_2 \rightarrow CO.$$

This exothermic reaction also contributes to maintaining the temperature in the gasifier.

Gasification Reaction

$$C + H_2O \rightarrow CO + H_2.$$

This endothermic reaction requires heat, which is provided by the exothermic reactions occurring simultaneously.

Water–Gas Shift Reaction

$$CO + H_2O \rightarrow CO_2 + H_2.$$

This reaction increases the hydrogen yield by converting carbon monoxide and steam into additional hydrogen and carbon dioxide. After gasification, the hot syngas produced is cooled and cleaned to remove impurities such as particulates, sulfur compounds, and other contaminants. The cleaned syngas, primarily composed of hydrogen, carbon monoxide, and carbon dioxide, undergoes the water–gas shift reaction to maximize hydrogen production. This reaction is typically conducted in two stages: a high-temperature shift (HTS) at about 350 °C and a low-temperature shift (LTS) at about 200 °C. Finally, the hydrogen is purified from the gas mixture using methods such as pressure swing adsorption (PSA). This physical separation process isolates high-purity hydrogen from other components like carbon dioxide and residual carbon monoxide, making it suitable for various applications. This comprehensive process leverages controlled chemical reactions and high-temperature conditions to efficiently convert coal into hydrogen, managing by-products and ensuring the purity of the final hydrogen output. Figure 3.6 illustrates steps to produce hydrogen from coal using coal gasification.

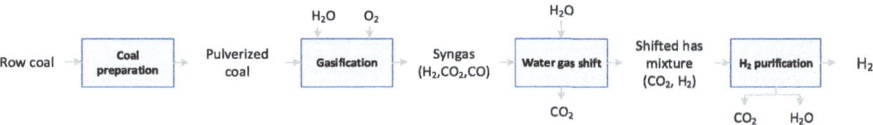

Fig. 3.6 Steps to produce hydrogen from coal using coal gasification

3.1.6 Electrolysis of Water

The production of hydrogen from water using electrolysis involves several key steps, each with specific inputs, outputs, and conditions. Initially, raw water is purified to remove impurities that could interfere with the electrolysis process. In the electrolysis step, purified water is subjected to an electric current in an electrolyzer, which splits the water molecules into hydrogen (H_2) and oxygen (O_2) gases. The process involves two primary reactions: at the anode, water is oxidized to produce oxygen gas, protons (H^+), and electrons (e^-); at the cathode, protons combine with electrons to form hydrogen gas. The overall reaction for water electrolysis can be summarized as:

$$2H_2O + \text{Electrical energy} \rightarrow 2H_2 + O_2.$$

The electrical energy required for electrolysis can come from various sources, including renewables such as solar, wind, hydroelectric, and geothermal energy, as well as non-renewable sources like natural gas, coal, and nuclear power. Using renewable energy sources enhances the sustainability and environmental benefits of the hydrogen production process. There are different types of electrolysis technologies, each with varying efficiencies, temperatures, and pressures. Table 3.1 provides a comparative overview of different electrolysis technologies used for hydrogen production, including alkaline electrolysis (AE), proton exchange membrane (PEM) electrolysis, and solid oxide electrolysis (SOE). The comparison highlights key parameters such as efficiency, operating temperature, pressure range, and notable characteristics of each technology. AE is cost-effective but less responsive to variable power; PEM is compact and suitable for renewables but costly, and SOE offers high efficiency but faces material durability challenges due to extreme temperatures.

After electrolysis, the produced gases (hydrogen and oxygen) are collected and separated. Additional purification steps may be necessary to achieve the required purity levels for specific applications, such as fuel cells or industrial processes. Figure 3.7 illustrates steps to produce hydrogen from electrolysis of water. This process efficiently produces high-purity hydrogen and oxygen gases from water using electrical energy. The choice of electrolysis technology and the source of electrical energy significantly influences the overall efficiency, sustainability, and cost-effectiveness of the hydrogen production process. By using renewable energy sources for electrolysis, the environmental impact can be minimized, making hydrogen a clean and versatile energy carrier.

3.1.7 Photobiological Processes

Hydrogen production through photobiological processes leverages the natural abilities of photosynthetic microorganisms, such as algae and cyanobacteria, to convert sunlight, water, and carbon dioxide into biomass and, under specific conditions,

3.1 Hydrogen Production

Table 3.1 Comparison of electrolysis technologies for hydrogen production

Type	Efficiency	Temperature Pressure	Notes	References
Alkaline electrolysis (AE)	60–70%	70–90 °C Atmospheric to 30 bar	AE uses an alkaline electrolyte solution, typically potassium hydroxide (KOH) or sodium hydroxide (NaOH). They are well-established and relatively inexpensive but have lower current densities and are less responsive to variable power sources	[1]
Proton exchange membrane (PEM)	60–70%	50–80 °C Up to 70 bar	PEM electrolyzers use a solid polymer electrolyte that conducts protons from the anode to the cathode. They are more compact, have higher current densities, and can operate at higher pressures, making them suitable for variable renewable energy sources. However, they are more expensive due to the use of precious metals as catalysts	[2]
Solid oxide electrolysis (SOE)	70–90%	700–1000 °C Typically atmospheric	SOE operates at high temperatures using a solid ceramic electrolyte. This high-temperature operation allows for higher efficiency and can utilize waste heat from industrial processes. However, the high operating temperatures pose challenges for materials and system durability	[3]

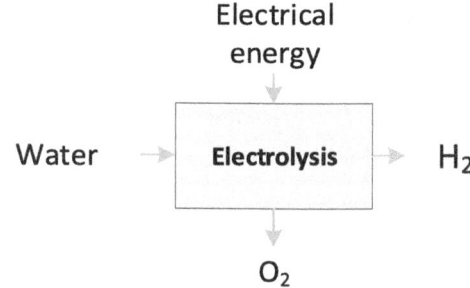

Fig. 3.7 Steps to produce hydrogen from electrolysis of water

hydrogen gas. The process begins with the cultivation of these microorganisms in bioreactors or open ponds. They use sunlight, water, carbon dioxide, and nutrients to perform photosynthesis, producing oxygen and organic compounds (glucose):

$$6CO_2 + 6H_2O + \text{light energy} \rightarrow C_6H_{12}O_6 + 6O_2.$$

Under anaerobic or nutrient-deprived conditions, these microorganisms switch to producing hydrogen gas via specific enzymatic pathways. Hydrogenases catalyze the reduction of protons to hydrogen gas:

$$2H^+ + 2e^- \rightarrow H_2.$$

The hydrogen gas produced by the microorganisms is collected, and various physical separation techniques are employed to purify the hydrogen from other gases, as illustrated in Fig. 3.8. This method of hydrogen production is environmentally friendly, leveraging renewable resources like sunlight and water, and can be sustainable if managed properly. The efficiency of hydrogen production depends on optimizing the growth conditions and genetic characteristics of the microorganisms involved.

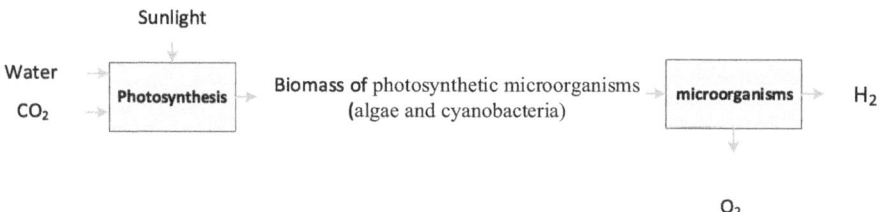

Fig. 3.8 Steps to produce hydrogen using photobiological processes

3.1 Hydrogen Production

Fig. 3.9 Steps to produce hydrogen using photoelectrochemical water splitting

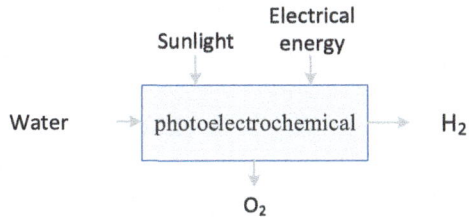

3.1.8 Photoelectrochemical Water Splitting

Hydrogen production through photoelectrochemical (PEC) water splitting involves using sunlight and semiconductor materials such as titanium dioxide (TiO_2) or gallium arsenide (GaAs) to drive the splitting of water molecules into hydrogen and oxygen gases. In the PEC cell, the semiconductor photoelectrode absorbs sunlight, generating electron–hole pairs that facilitate the redox reactions needed to split water. At the photoanode, water is oxidized to produce oxygen gas, protons, and electrons:

$$2H_2O \rightarrow O_2 + 4H^+ + 4e^-.$$

These electrons travel through an external circuit to the cathode, where they combine with protons to form hydrogen gas:

$$4H^+ + 4e^- \rightarrow 2H_2.$$

The overall reaction for PEC water splitting is:

$$2H_2O + \text{sunlight} \rightarrow 2H_2 + O_2.$$

This process generally occurs at ambient temperature (20–25 °C) and atmospheric pressure (1 atm). The hydrogen and oxygen gases produced are then collected and separated, as illustrated in Fig. 3.9. Purification techniques such as gas–liquid separation, pressure swing adsorption (PSA), or membrane separation may be used to ensure the hydrogen gas reaches the required purity levels for its intended applications. This method leverages renewable solar energy to produce clean hydrogen fuel, contributing to sustainable energy solutions.

3.1.9 Thermochemical Water Splitting

There are several thermochemical water splitting cycles with various steps such as Copper-Chlorine, Magnesium-Chlorine and Sulfur-Iodine. As an example, hydrogen production through thermochemical water splitting using the Sulfur-Iodine (S-I) cycle involves a series of chemical reactions driven by high temperatures. The process begins with the decomposition of sulfuric acid (H_2SO_4) at approximately 850–950 °C, resulting in the formation of water (H_2O), sulfur dioxide (SO_2), and oxygen (O_2):

$$H_2SO_4 \rightarrow H_2O + SO_2 + \frac{1}{2}O_2.$$

Next, the sulfur dioxide (SO_2) reacts with iodine (I_2) and water (H_2O) at lower temperatures (120–150 °C) to produce hydrogen iodide (HI) and regenerate sulfuric acid (H_2SO_4):

$$SO_2 + I_2 + 2H_2O \rightarrow 2HI + H_2SO_4.$$

In the final step, hydrogen iodide (HI) is decomposed at high temperatures (300–450 °C) to yield hydrogen (H_2) and iodine (I_2):

$$2HI \rightarrow H_2 + I_2.$$

The iodine is then recycled back into the second step, making the process cyclic. This thermochemical cycle efficiently converts water into hydrogen and oxygen by utilizing high-temperature heat, which can be sourced from concentrated solar power, nuclear reactors, or other high-temperature heat sources. The overall process is designed to be sustainable and can leverage renewable or nuclear energy to produce clean hydrogen fuel.

3.1.10 Biomass Gasification

Hydrogen production from biomass involves several key steps. First, raw biomass is prepared by drying and grinding it to an appropriate size. The processed biomass is then subjected to gasification, where it is heated in the presence of controlled amounts of oxygen and steam, producing syngas, a mixture of hydrogen, carbon monoxide, carbon dioxide, and methane, along with by-products like tar, char, and ash. The main reactions during gasification include combustion, partial oxidation, steam gasification, and methanation.

The primary reactions involved in the gasification process are:

Combustion Reaction

$$C + O_2 \rightarrow CO_2.$$

This exothermic reaction generates heat, which helps maintain the high temperature required for other reactions.

Partial Oxidation Reaction

$$C + \frac{1}{2}O_2 \rightarrow CO.$$

This exothermic reaction also contributes to maintaining the temperature in the gasifier.

Gasification Reaction

$$C + H_2O \rightarrow CO + H_2.$$

This endothermic reaction requires heat, which is provided by the exothermic reactions occurring simultaneously.

Water–Gas Shift Reaction

$$CO + H_2O \rightarrow CO_2 + H_2.$$

Methanation

$$CO_2 + 4H_2 \rightarrow CH_4 + 2H_2O.$$

The raw syngas is then cleaned to remove impurities such as tar, particulates, and sulfur compounds. This cleaned syngas undergoes the water–gas shift reaction, where carbon monoxide reacts with steam to produce additional hydrogen and carbon dioxide. This reaction is typically carried out in two stages, high-temperature shift (HTS) and low-temperature shift (LTS). Finally, the hydrogen is purified from the shifted gas mixture using techniques like pressure swing adsorption (PSA) or membrane separation. This results in high-purity hydrogen suitable for various applications.

3.2 Hydrogen Storage

Hydrogen, a clean and versatile energy carrier, is essential in the transition to a sustainable energy future. Ensuring the safe and efficient storage of hydrogen energy is one of the primary obstacles to its widespread adoption. Physical storage, chemical storage, and materials-based storage are the three primary categories into which hydrogen storage technologies can be broadly classified. In order to facilitate the practical and widespread use of hydrogen, each type of hydrogen presents distinct advantages and challenges that must be resolved. Figure 3.10 shows an overview of hydrogen storage technologies, categorized into three main types: **physical storage, chemical storage**, and **materials-based storage**. Physical storage includes compressed hydrogen gas and liquid hydrogen, while chemical storage encompasses metal hydrides, chemical hydrogen storage, and chemical hydrides such

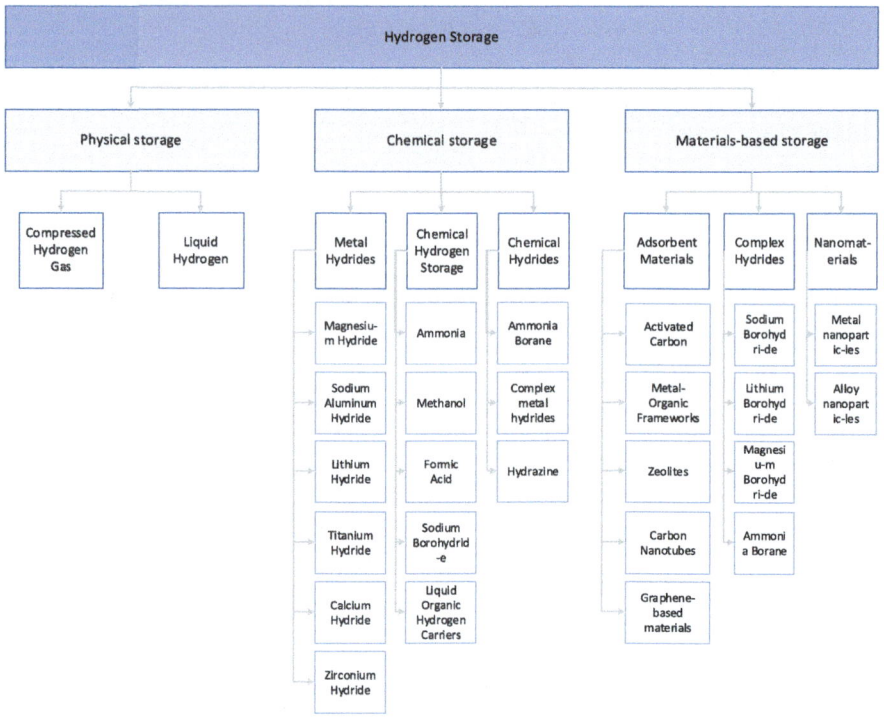

Fig. 3.10 Hydrogen storage methods and classification

as ammonia borane and formic acid. Materials-based storage is further divided into adsorbent materials, complex hydrides, and nanomaterials, which utilize advanced materials like metal–organic frameworks (MOFs), carbon nanotubes, and sodium borohydride for efficient hydrogen storage.

3.2.1 Physical Storage

Hydrogen storage techniques predominantly entail preserving the element in its elemental state—either as a compressed gas or as a cryogenically liquefied liquid. These methods are essential in applications from fuel cell vehicles to aerospace, where storage efficiency, safety, and energy requirements must be meticulously balanced. The two primary physical storage methods each offer distinct benefits and drawbacks that warrant careful analysis.

Storage of compressed hydrogen gas depends on high-pressure containment systems, generally functioning within a range of 350–700 bar (5000–10,000 psi). This established method boasts a high gravimetric density, allowing for substantial hydrogen mass storage relative to the storage system's weight—typically around

50 kg/m^3. The developed infrastructure for compressed gas storage enhances its incorporation into current technologies, especially in fuel cell vehicles and stationary energy storage systems. Nonetheless, the low volumetric density of compressed hydrogen requires larger storage capacities, and the high pressures involved necessitate durable, frequently substantial, containment solutions like composite overwrapped pressure vessels (COPVs). Moreover, the intrinsic safety hazards linked to high-pressure gas storage and the substantial energy demand—around 6 kWh/kg for compression to 700 bar—present considerable technical and economic obstacles that need to be resolved for wider implementation [4].

Conversely, liquid hydrogen storage utilizes cryogenic methods to convert hydrogen gas into a liquid state at around − 253 °C. This method is especially beneficial in situations where high-energy density and compact storage are critical, such as in aerospace applications and large-scale transportation. Liquid hydrogen demonstrates superior gravimetric and volumetric densities, facilitating more efficient spatial utilization than compressed gas storage. Nonetheless, the cryogenic storage method presents its own array of complexities. Preserving the requisite extremely low temperatures for liquefaction necessitates energy-intensive cooling systems and superior insulation, which elevate operational costs and complicate the design of the storage system. Furthermore, the unavoidable boil-off losses—stemming from gradual evaporation over time—can result in diminished storage efficiency, and the liquefaction process is energy-intensive, necessitating between 10 and 13 kWh/kg. These factors emphasize the imperative for ongoing research and technological progress to enhance the efficiency and safety of liquid hydrogen storage systems [5].

In conclusion, although physical storage methods for hydrogen present feasible options for energy storage across diverse applications, each method entails a complicated balance of advantages and disadvantages. Compressed hydrogen gas storage leverages established technology and infrastructure; however, it is constrained by low volumetric density and elevated energy costs associated with compression. In contrast, liquid hydrogen storage offers enhanced density and compactness but necessitates greater energy for liquefaction and poses significant cryogenic management difficulties. Overcoming these challenges via advancements in materials science, system design, and energy efficiency will be essential for the future development and practical application of hydrogen storage technologies in the changing energy landscape.

3.2.2 Chemical Storage

Chemical hydrogen storage methods present a viable approach for the secure and efficient storage of hydrogen by integrating it into chemical compounds, allowing for hydrogen release via controlled chemical reactions. This method differs from physical storage techniques—like compressed gas or cryogenic liquid storage—by potentially providing greater energy densities and enhanced safety characteristics.

Chemical storage strategies are primarily classified according to the characteristics of the compounds employed and the processes through which hydrogen is liberated.

A notable category within this field is metal hydrides. Metal hydrides are compounds capable of absorbing hydrogen gas and storing it in a solid form, subsequently releasing the hydrogen when heated. Their elevated volumetric density facilitates the storage of a significant amount of hydrogen in a relatively confined space, which is especially beneficial for applications with stringent spatial limitations. The requirement to apply heat for hydrogen release presents challenges to efficiency and operational practicality, particularly in situations necessitating rapid or energy-efficient hydrogen extraction. The intrinsic weight of numerous metal hydrides may limit their applicability in mobile or weight-sensitive contexts, highlighting a trade-off between energy density and system mass.

A further promising strategy involves chemical hydrogen storage utilizing liquid or solid compounds, including ammonia, methanol, and liquid organic hydrogen carriers (LOHCs). In these systems, catalysts are frequently utilized to promote the liberation of hydrogen via particular chemical reactions. In comparison to traditional storage methods such as compressed or liquefied hydrogen, these chemical techniques can attain superior energy densities and provide improved safety regarding handling and storage. However, dependence on chemical processing to liberate the stored hydrogen presents challenges related to reaction kinetics, catalyst efficacy, and the comprehensive energy balance of the system. These factors pose significant challenges that must be overcome to achieve the full potential of chemical hydrogen storage technologies.

A specific category of chemical storage encompasses chemical hydrides—solid compounds that liberate hydrogen through hydrolysis or other chemical reactions. The principal benefit of chemical hydrides is their capacity to store hydrogen at comparatively low pressures and temperatures, enhancing safety in handling and facilitating storage infrastructure. Nonetheless, the hydrogen release mechanism in these systems generally necessitates the addition of water or another reactant, resulting in the production of by-products that require management. This supplementary requirement complicates system design and may affect the overall efficiency of the storage and release process, necessitating further investigation into optimized reaction pathways and by-product remediation methods.

In conclusion, although chemical storage methods for hydrogen offer considerable benefits regarding energy density and safety relative to physical storage options, each method—be it metal hydrides, chemical hydrogen storage compounds, or chemical hydrides—presents distinct technical challenges. It is essential to tackle challenges associated with thermal regulation, reaction efficacy, and by-product management to progress these technologies toward viable, scalable implementations in the burgeoning hydrogen economy.

3.2.3 Materials-Based Storage

Materials-based hydrogen storage is a dynamic and promising research domain that utilizes advanced materials to adsorb or absorb hydrogen on the surface or within the material's structure. This method utilizes the distinctive physical and chemical characteristics of these materials to facilitate storage under comparatively mild conditions relative to traditional physical storage techniques. The adaptability of materials-based storage is apparent in the diverse strategies being explored, such as the utilization of adsorbent materials, complex hydrides, and nanomaterials.

Adsorbent materials, including activated carbon, metal–organic frameworks (MOFs), and zeolites, possess highly porous structures and large surface areas that enhance the adsorption of hydrogen molecules. These materials provide substantial safety advantages by facilitating hydrogen storage at comparatively low pressures and temperatures, thereby mitigating the hazards linked to high-pressure systems. Nonetheless, despite their benefits in safety and operational efficiency, the hydrogen storage capacity of these adsorbents typically falls short compared to other methods. Current research seeks to augment the adsorption properties of these materials, potentially by altering pore structures and surface chemistry to enhance storage efficacy.

Besides adsorbents, complex hydrides offer an alternative method for chemical hydrogen storage. In these systems, hydrogen is chemically integrated into the structures of compounds like sodium borohydride and lithium borohydride. These intricate hydrides can retain a significant amount of hydrogen, which can be liberated via regulated chemical reactions or by the application of heat. The substantial storage capacity of complex hydrides is appealing; however, their practical use is frequently impeded by the requirement for specific temperature and pressure conditions to trigger hydrogen release. The reliance on particular reaction environments poses a challenge in creating efficient and user-friendly storage systems, requiring additional research to optimize reaction pathways and system integration.

Nanomaterials provide an extra dimension of potential improvement in hydrogen storage through the application of nanotechnology principles. Metal and alloy nanoparticles exhibit exceptionally high surface area-to-volume ratios, which can markedly enhance the kinetics of hydrogen adsorption and desorption. The integration of nanostructured materials can facilitate accelerated hydrogen absorption and desorption, thereby mitigating certain kinetic constraints evident in bulk materials. Research on the use of nanomaterials for hydrogen storage remains in its nascent phase; however, preliminary investigations indicate that these materials may significantly enhance storage capacity and efficiency, thereby expanding the range of materials-based hydrogen storage options.

In conclusion, materials-based hydrogen storage provides a comprehensive method for hydrogen storage by leveraging the distinctive properties of advanced adsorbent materials, complex hydrides, and nanomaterials. Each strategy has distinct advantages and challenges: adsorbents provide safety and operational benefits but may necessitate capacity enhancements; complex hydrides offer significant storage potential but are limited by particular release conditions; and nanomaterials promise improved kinetics and efficiency yet are still under active research. Ongoing interdisciplinary research in materials science, chemistry, and engineering will be essential for addressing these challenges and progressing the development of efficient and practical hydrogen storage technologies.

3.3 Hydrogen Transport

Modes of hydrogen transportation can be classified based on the form of hydrogen, which can be either gaseous or liquid. Gaseous hydrogen can be transported via compressed gas trailers, tube trailers, or pipelines. Compressed gas trailers employ trailers with high-pressure cylinders to store hydrogen at pressures ranging from 200 to 500 bar (2900–7250 psi). This method is particularly suitable for transporting smaller quantities of hydrogen over shorter distances, making it a popular option for localized deliveries and fueling stations. Nevertheless, the use of specialized handling equipment and strict adherence to safety protocols is required due to the high pressures involved. Research suggests that compressed gas trailers have a maximum capacity of approximately 500 kg of hydrogen per trailer, which restricts their suitability for widespread distribution of hydrogen on a large scale. Tube trailers utilize clusters of high-pressure seamless steel tubes affixed to trailers, enabling the storage of hydrogen at significantly elevated pressures, up to 800 bar (11,600 psi). This technology provides a greater capacity in comparison to compressed gas trailers, rendering them more appropriate for transportation requirements on a larger scale. Nevertheless, the utilization of tube trailers necessitates rigorous safety protocols and specialized infrastructure for the handling and unloading processes, owing to the high pressures involved. Studies indicate that tube trailers have the capacity to transport up to 1000 kg of hydrogen per trailer, thereby increasing their capability for delivering large quantities of hydrogen. Pipelines are a system of connected pipes specifically designed for transporting gaseous hydrogen over long distances. These pipelines, made of durable materials such as steel or composites, are engineered to endure high pressures and prevent leaks, guaranteeing the secure and effective transportation of hydrogen across long distances. Although pipelines necessitate substantial upfront investment and continuous upkeep, they provide unparalleled benefits in terms of efficiency and capacity for the distribution of hydrogen on a large scale. Research indicates that pipelines have the capacity to transport hydrogen at rates surpassing 10,000 kg per hour, making them an essential element of forthcoming hydrogen infrastructure [6].

Hydrogen can also be transported in a liquid form via cryogenic tankers or cryogenic tank trucks. Cryogenic tankers are specialized ships designed with well-insulated cryogenic tanks to keep hydrogen in its liquid form at an extremely frigid temperature of − 253 °C. This technique facilitates the conveyance of substantial amounts of hydrogen over vast bodies of water, presenting a promising resolution for extensive, long-distance hydrogen commerce. Nevertheless, the existing quantity and capability of cryogenic tankers are restricted, and the technology necessitates substantial investment and specialized handling due to the difficulties linked to cryogenic liquids. However, the progress in cryogenic technologies and increasing global demand for hydrogen indicate a positive outlook for cryogenic tankers. Cryogenic tank trucks are specifically designed vehicles equipped with highly insulated cryogenic tanks for the purpose of transporting liquid hydrogen on roadways. They provide a flexible solution for delivering goods to specific locations and allow access to areas that do not have pipeline infrastructure. Nevertheless, like cryogenic tankers, they necessitate meticulous handling and specialized equipment owing to the exceedingly low temperatures involved. In addition, cryogenic tank trucks have a relatively smaller capacity compared to other transportation methods, which restricts their suitability for distributing hydrogen on a large scale.

Other transportation methods of hydrogen are hydrogen carriers, such as ammonia, methanol, or metal hydrides. These carriers provide an alternative method for transporting hydrogen by chemically storing it within other substances. Conventional transportation methods can be used to transport these carriers, which may provide benefits in terms of safety and ease of handling when compared to gaseous or liquid hydrogen. Nevertheless, the utilization of hydrogen necessitates the establishment of supplementary infrastructure and procedures for the extraction of hydrogen at the intended location, resulting in energy inefficiencies and heightened intricacy. The selection of an appropriate carrier is contingent upon several factors, such as the intended use, distance of transportation, and the presence of conversion infrastructure. Table 3.2 provides an overview of hydrogen transportation modes, detailing their advantages and disadvantages. Additionally, choosing the most suitable method of transportation for hydrogen requires a thorough assessment of several factors, such as distance, volume, available infrastructure, cost considerations, and safety and environmental considerations. Efficient and affordable hydrogen transportation solutions are still a significant obstacle in fully harnessing the potential of hydrogen as a clean energy carrier. Continuing research and innovation in this field are anticipated to result in progress in transportation technologies and infrastructure, thereby facilitating the development of a future powered by hydrogen.

Table 3.2 Overview of transportation modes for hydrogen: advantages and disadvantages

Mode of transportation		Advantages	Disadvantages
Gaseous	Compressed gas trailers	Suitable for smaller volumes and shorter distances, flexible	Lower capacity, higher cost per unit, requires specialized handling equipment
	Tube trailers	Higher capacity than compressed gas trailers, suitable for larger-scale transportation	Requires specialized infrastructure and handling, safety concerns due to high pressures
	Pipelines	Cost-effective for long distances, high capacity, minimal environmental impact during operation	High initial investment, potential for leaks and land disruption, limited existing infrastructure
Liquid	Cryogenic tankers	High capacity, suitable for long distances, potential for lower cost per unit in the future	High initial investment, limited existing infrastructure, requires specialized handling and safety measures
	Cryogenic tank trucks	Flexible for localized deliveries, suitable for areas without pipeline access	Lower capacity, higher cost per unit, requires specialized handling and safety measures
Others	Hydrogen carriers	Potentially safer and easier to handle than gaseous or liquid hydrogen	Requires additional infrastructure and processes for hydrogen extraction, energy losses during conversion

References

1. El-Shafie M (2023) Hydrogen production by water electrolysis technologies: a review. Results Eng 20:101426. https://doi.org/10.1016/j.rineng.2023.101426
2. Shiva Kumar S, Himabindu V (2019) Hydrogen production by PEM water electrolysis—a review. Mater Sci Energy Technol 2:442–454. https://doi.org/10.1016/j.mset.2019.03.002
3. Mueller M, Klinsmann M, Sauter U, Njodzefon JC, Weber A (2024) High temperature solid oxide electrolysis—technology and modeling. Chem-Ing-Tech 96:143–166. https://doi.org/10.1002/cite.202300137
4. Patonia A, Poudineh R (2023) Hydrogen storage for a net-zero carbon future. Oxford Inst Energy Stud
5. Goldmeer J, Catillaz J (2022) Hydrogen for power generation. Gea34805 (03/22):17
6. Muhammed NS, Gbadamosi AO, Epelle EI, Abdulrasheed AA, Haq B, Patil S et al (2023) Hydrogen production, transportation, utilization, and storage: recent advances towards sustainable energy. J Energy Storage 73. https://doi.org/10.1016/j.est.2023.109207

References

Open Access This chapter is licensed under the terms of the Creative Commons Attribution 4.0 International License (http://creativecommons.org/licenses/by/4.0/), which permits use, sharing, adaptation, distribution and reproduction in any medium or format, as long as you give appropriate credit to the original author(s) and the source, provide a link to the Creative Commons license and indicate if changes were made.

The images or other third party material in this chapter are included in the chapter's Creative Commons license, unless indicated otherwise in a credit line to the material. If material is not included in the chapter's Creative Commons license and your intended use is not permitted by statutory regulation or exceeds the permitted use, you will need to obtain permission directly from the copyright holder.

Chapter 4
Ammonia

4.1 Ammonia Production

Consisting of one nitrogen atom and three hydrogen atoms, ammonia (NH_3) is conventionally produced using various methods. Some of these are more industrially suitable than others, depending on raw material availability, cost, and technology [1]. Largely for economic reasons, the synthetic hydrogen feedstock for ammonia is produced almost entirely from fossil fuels; natural gas accounts for 72%; coal accounts for 22%; heavy oil and naphtha account for around 4% and 1%, respectively, while the remaining 1% is derived from other feedstocks [2]. Figure 4.1 categorizes hydrogen production methods based on non-renewable sources (natural gas, oil, and coal) and renewable sources (water, sunlight, heat, and biomass). Non-renewable pathways include steam methane reforming, methane pyrolysis, partial oxidation, plasma reforming, and gasification, whereas renewable methods consist of electrolysis, photobiological, photoelectrochemical, thermochemical, and biomass gasification processes. The hydrogen (H_2) produced from these sources is combined with nitrogen (N_2) to synthesize ammonia (NH_3), a crucial compound for fertilizers, energy storage, and industrial applications.

4.1.1 The Haber-Bosch Process

Developed by Fritz Haber and Carl Bosch more than 100 years ago, the Haber-Bosch process is the most widely adopted method to produce ammonia, shaping about 85% of the global production [3]. Figure 4.2 illustrates the steam methane reforming (SMR) pathway for producing hydrogen (H_2) as a feedstock for ammonia (NH_3) synthesis. The process begins with methane (CH_4) and water (H_2O) undergoing reforming to produce a mixture of carbon monoxide (CO), carbon dioxide (CO_2), and hydrogen (H_2). The water-gas shift reaction then converts CO into additional H_2

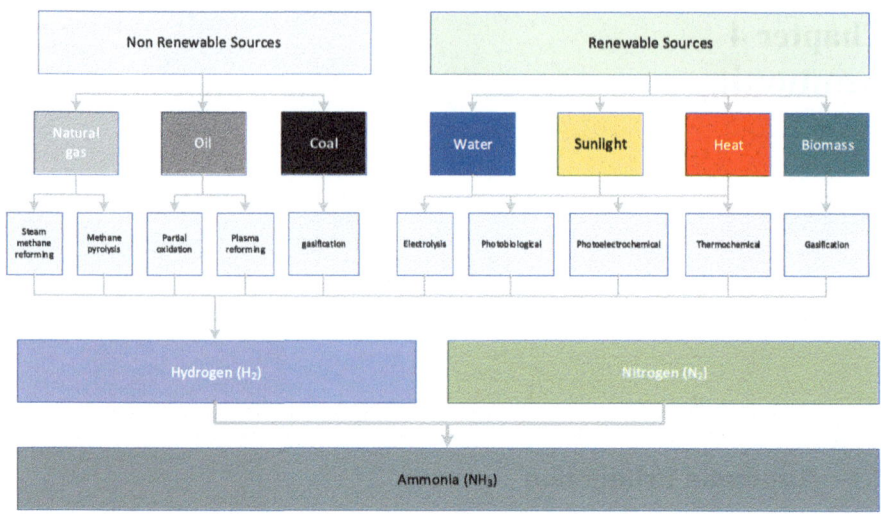

Fig. 4.1 Hydrogen and nitrogen production pathways for ammonia (NH$_3$) synthesis

and CO$_2$. Gas removal eliminates CO$_2$, leaving purified H$_2$, which is combined with nitrogen (N$_2$) in the Haber-Bosch reactor to synthesize ammonia (NH$_3$), a crucial compound for fertilizers, energy storage, and industrial applications.

$$3H_2 + N_2 \rightleftharpoons 2NH_3 \quad \Delta H°_{25°C} = -46.2\,\text{kJ/mol}.$$

As shown in the chemical equation, ammonia synthesis is an exothermic reaction (i.e., $\Delta H < 0$) occurring spontaneously at low temperatures. While the reaction is favored at room temperature, the reaction rate at ambient conditions is too slow to be applicable for commercial practices. Thus, high pressure and temperature are often required to prompt the kinetics of the reaction into a desirable conversion rate. At temperature and pressure ranges of 400–500 °C and 10–30 MPa, respectively, and the presence of an iron-based catalyst, these conditions are demanded to effectively synthesize ammonia due to the high dissociation energy (941 kJ/mol) of nitrogen triple bond (N ≡ N).

Ammonia production from natural gas is conducted by reacting methane (CH$_4$) with steam (i.e., SMR) and air coupled with subsequent stages of water and CO$_2$ removal. The nitrogen and hydrogen are then compressed to a relatively high pressure to be fed to the synthetic reactor, where the iron-based catalyst is immersed

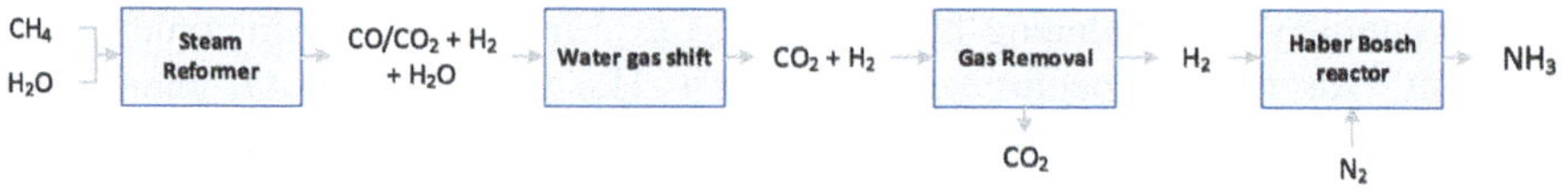

Fig. 4.2 Steam methane reforming (SMR) process for ammonia (NH$_3$) production

4.1 Ammonia Production

inside. Along with other unreacted containments such as hydrogen, argon, and other impurities, the produced ammonia is condensed in order to separate the ammonia from the other gases. The unreacted hydrogen and nitrogen are recycled via a bypass stream and mixed with the new feedstock. To avoid accumulation of impurities, especially inert gases (e.g., argon), a small amount of the gases is purged from the process. The ammonia synthetic reactor releases some amount of heat as a consequence of the exothermic formation of ammonia, which can be recovered and integrated with other processes such as steam and power generation to cut out costs.

4.1.2 Electrochemical Ammonia Synthesis

Electrochemical ammonia synthesis (EAS) is an attractive method for ammonia production since it is usually conducted using abundant nitrogen (N_2) and water (H_2O) near-ambient conditions, where electricity is the driving force. When compared to the traditional Haber-Bosch process, this method consumes approximately 20% less energy thanks to its simplicity suggesting minimal system configuration, reduced control complexity, and lower investment costs [4]. The modern significance of EAS lays in leveraging renewable electricity to decarbonize ammonia production, enabling decentralization and sustainability (e.g., on-site production for local fertilizers and fuel generation without the need for massive industrial plants). Figure 4.3 illustrates the green ammonia production process, where hydrogen (H_2) is generated from water (H_2O) using an electrolyzer powered by renewable electricity. Oxygen (O_2) is separated as a by-product, while nitrogen (N_2) is obtained from air through an O_2 and H_2O removal process. The purified H_2 and N_2 are then compressed before entering the ammonia synthesis reactor, where they undergo the Haber-Bosch process to form ammonia (NH_3). This method is considered a sustainable alternative to conventional ammonia production, as it relies on renewable power sources and does not emit carbon dioxide (CO_2).

An electrochemical cell consists of three primary components: the electrolyte, anode, and cathode. The electrolyte, an ion-conducting medium, separates the electrodes while enabling ion transport between them. Oxidation occurs at the anode,

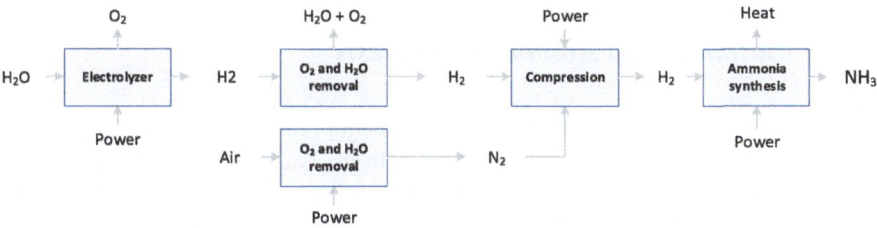

Fig. 4.3 Ammonia production via electrolysis

and reduction takes place at the cathode. The electrodes are linked by an external circuit that facilitates electron flow, completing the circuit.

In a typical electrochemical cell, N_2 gas is introduced into the cathode chamber, where it undergoes reduction to form ammonia (NH_3). Simultaneously, water (H_2O) at the anode is oxidized, producing oxygen gas. This process allows ammonia synthesis under mild conditions, offering a sustainable alternative to conventional methods.

Overall reaction:

$$N_2 + 3H_2O \rightarrow 2NH_3 + 1.5O_2.$$

In the absence of strong and selective catalytic electrodes, however, the reaction is almost unachievable. This is because, at the anode, nitrogen (N_2) is a highly stable inert that requires to be stripped from the triple nitrogen bond ($N \equiv N$) to react with the proton (H^+) that is detached from the water hydrogen bond [4]. Therefore, a N_2 reduction catalyst and an O_2 evolution catalyst are applied at the cathode and anode, respectively, to minimize overpotentials at both electrodes. The catalysts play a significant role in overcoming the strong triple bond and providing the facilitation to receive electrons from the external circuit and interact with nitrogen to form NH_3 in a stepwise hydrogenation process.

A number of studies have investigated different combinations of electrolytes and electrode materials, and they could be summarized into two main categories, depending on the operating temperature of the electrochemical cell: low temperature ($T < 100\,°C$) and high temperature ($400 < T < 700\,°C$). Table 4.1 summarizes key studies on electrochemical nitrogen reduction for ammonia (NH_3) synthesis, highlighting differences in temperature range, electrolyte composition, electrode materials, reaction rates, and efficiency. The studies explore various electrochemical systems, including room-temperature Nafion-based approaches, molten salt systems, proton-conducting ceramics, and composite electrolytes. Efficiency and reaction rates vary significantly, with some studies achieving high efficiency but low reaction rates, while others propose novel nitride transport mechanisms. Notably, some reported efficiencies exceed 100%, raising concerns about data validation and reaction yield limitations.

4.1.3 Plasma-Based Ammonia Synthesis

Plasma is an ionized gas formed when a significant portion of atoms and molecules have been stripped or excited by gaining electrons, resulting in a mixture of free electrons, ions, and neutral particles, often referred to as photons [12]. This highly energetic state of matter is distinguished by its ability to generate reactive species that drive chemical reactions under conditions unattainable by conventional means. In the case of ammonia synthesis, plasma generates highly reactive species which facilitate N_2 fixation, can be operated under atmospheric pressure, and can be powered with

4.1 Ammonia Production

Table 4.1 Overview of electrochemical nitrogen reduction studies for ammonia synthesis

Study	Temperature range	Electrolyte	Electrodes	Rxn rate $\left[\frac{mol}{s\,cm^2}\right]$	Efficiency	Notes
[5]	Room temperature	Nafion (H^+ form, converted to NH_4^+ form)	Pt/C	1.14×10^{-9}	< 1%	First synthesis from air and water; Pt not ideal for N_2 reduction
[6]	300–400 °C	Molten salts (LiCl-KCl-CsCl)	Nitrides	$\sim 10^{-8}$	–	Unique N^{3-} transport mechanism; innovative but limited data
[7]	300–400 °C	Molten salt (LiCl-KCl-CsCl)	Not specified	$\sim 10^{-8}$	–	Patented process using molten salt-based nitride transport
[8]	450–700 °C	$SrCe_{0.95}Yb_{0.05}O_{3-\delta}$ (proton)	Pd	10^{-12} to 10^{-11}	90%	High efficiency with H_2 and N_2; low rate with steam and N_2.
[9]	450–700 °C	$SrCe_{0.95}Yb_{0.05}O_{3-\delta}$ or $ZrY_{0.92}Y_{0.08}O_{2-\delta}$	Ru (on Ag or Pd)	10^{-12} to 10^{-11}	~ 0.01%	Formation rate ~ 3 orders lower than Marnellos; NH_3 < 1 ppm at low rate

(continued)

Table 4.1 (continued)

Study	Temperature range	Electrolyte	Electrodes	Rxn rate $\left[\frac{mol}{s\,cm^2}\right]$	Efficiency	Notes
[10]	400–800 °C	$Ce_{0.8}M_{0.2}O_{2-\delta}$ (M = La, Y, Gd, Sm)	Ag-Pd	$2\text{–}7 \times 10^{-9}$	>100%	Claims unrealistic efficiency; proton source limits NH_3 yield
[11]	400–450 °C	Carbonate-oxide composite	$La_{0.6}Sr_{0.4}Fe_{0.8}Cu_{0.2}O_3\text{–}\delta\text{–}CeO_2$	5.39×10^{-9}	~40 to 50%	Composite electrolytes; reasonable rates with intermediate efficiency

4.1 Ammonia Production

renewable electricity, which makes it perfectly suited for decentralized and intermittent production. The recent advances in employing plasma-assisted process involve plasma electrochemical and plasma catalysis used to favor the desired reaction.

As N_2 dissociation is usually the bottleneck of ammonia production (i.e., the main reason for the sluggish kinetics), plasma could overcome this obstacle by striking the N_2 molecules with an electron impact, ionization, and excitation, creating N atoms, ions, and excited species, which easily react with other compounds such as NH_3. However, plasma catalytic reduction of N_2 faces major challenges. While it could easily activate the relatively inert N_2 molecule, it only excites the light electrons on the surface yielding low concentration of ammonia at higher energy consumption levels than the traditional method. To put to scale, a plasma process with a relatively low energy consumption of 2 MJ mol^{-1}, being close to that of Haber-Bosch (0.52–0.81 MJ mol^{-1}), yields less than 0.1 vol% of diluted ammonia. As a result, the recovery from a very diluted ammonia product will be very challenging and highly energy-intensive.

A Plasma Nitrogen Oxidation and Catalytic Reduction to Ammonia (PNOCRA) process was proposed by Halloevoet and his colleagues, in 2020, witnessing that plasma favors oxidation reactions rather than reductions [13]. Via combining plasma with engine exhaust gas after-treatment technologies to overcome the inefficiency of plasma processes for ammonia synthesis, it was suggested that N_2 is first oxidized to NO_x and then reduced to NH_3.

The process, inspired by the Birkeland-Eyde method and automotive industry technologies, employs an electric arc to oxidize nitrogen (N_2) from air into nitric oxide (NO) [14]. A strong magnetic field is applied to spread the arc into a disk shape with an approximate diameter of 1.8 m, enhancing the interaction between air and the arc. As air passes through the arc, nitrogen reacts with oxygen at temperatures around 1100 °C, resulting in a gas mixture containing 1–2% NO.

After that, the gas enters a large oxidation chamber, where NO gradually oxidizes to NO_2. The gas mixture cools from 200 to 50 °C in cooling towers, improving the absorption rate. The cooled gas is brought into contact with water, forming nitric acid (HNO_3) through the reaction:

$$3NO_2 + H_2O \rightarrow 2HNO_3 + NO.$$

The process consumes approximately 2.4 MJ per mol of NO produced, reflecting the energy-intensive nature of arc-based nitrogen oxidation.

A portion of NO is re-released during this reaction, requiring continuous oxidation and reoxidation cycles until nearly all NO is converted. This step yields approximately 30% concentrated nitric acid.

In the next phase, nitric oxides (NO_x) are selectively reduced to ammonia (NH_3) through a catalytic hydrogenation process. However, the gas mixture from the oxidation chamber contains unreacted oxygen, which could consume hydrogen during ammonia synthesis. To prevent this, the process incorporates a NO_x trap—a system originally developed for automotive exhaust treatment. This trap, often made of barium oxide (BaO) on alumina with finely dispersed platinum, selectively adsorbs NO_x while allowing oxygen to pass.

Once the NO_x is adsorbed, it reacts with hydrogen in a reducing environment:

$$Ba(NO_3)_2 + 8H_2 \rightarrow 2NH_3 + BaO + 5H_2O.$$

Alternatively, under slightly different conditions, the trap can convert NO_x directly to nitrogen:

$$Ba(NO_3)_2 + 5H_2 \rightarrow N_2 + BaO + 5H_2O.$$

The process achieves selective ammonia production by controlling the catalyst's environment and reaction conditions, forming ammonia efficiently even in the presence of oxygen. PNOCRA method offers a decentralized, energy-efficient pathway for ammonia synthesis powered by renewable electricity.

While non-conventional technologies typically cannot compete with Haber-Bosch at large scale, alternative technologies such as plasma-based ammonia production can be beneficial for small-scale production. This is important for intermittent operations required in the case of rapid variation in the availability of solar and wind power.

4.1.4 Biological Ammonia Synthesis

Biological approaches are considered eco-friendly as they are natural processes that do not produce harmful by-products. Biological nitrogen fixation, nitrification, nitrate/nitrate reduction, urea hydrolysis, metabolic engineering of microorganisms, and in vitro ruminal microbial fermentation of protein biomass, are all alternative and modern approaches for biological ammonia production that could operate near-ambient conditions. The most widely reported methods, however, are biological nitrogen fixation (BNF) and metabolic engineering of microorganisms. Another approach is biological ammonia production by rumen bacteria which is relatively new in the field albeit has shown potential to complement the bioproduction.

BNF is a natural process that converts atmospheric molecular N_2 to NH_3 by a reaction, reliant on adenosine-triphosphate (ATP) reduction catalyzed by the nitrogenase enzyme. It is responsible for more than a half of the bioavailable N_2 that sustains all life forms.

$$N_2 + 8H^+ + 8e^- + 16ATP \rightarrow 2NH_3 + H_2 + 16ADP + 16P_i.$$

Typically, the reaction takes place at the active site, the FeMo cofactor (FeMoco), which is a metal sulfur cluster with the stoichiometry [Mo-7Fe-9S-N]. The biochemical equation highlights several notable aspects of biological ammonia synthesis. One key point is that nitrogenase utilizes protons and electrons (e^-), rather than molecular hydrogen (H_2). The protons are sourced from the surrounding solution, while the electrons are delivered by another iron-sulfur cluster within the enzyme. This electron transfer allows electrons to be introduced with a specific chemical potential, supplying energy to drive the reaction forward. The energy needed to overcome the reaction barrier is derived from the hydrolysis of ATP and water, producing adenosine diphosphate (ADP) and inorganic phosphate (HPO_4^{2-}). This mechanism reflects a common biological strategy for storing and transferring energy.

The major challenge with this research effort is that eight molecules of ATP to produce a molecule of ammonia is an anoxic condition. Researchers are actively experimenting to mimic the natural process of BNF by isolating nitrogen-fixing bacteria and nitrogenase for synthetic ammonia production.

Such research involves the use of polysaccharide membranes to protect nitrogenase from oxygen exposure. The high energy requirement of nitrogenase catalysis nonetheless persists, making its reaction rate slower than most enzymes in nature.

4.1.5 Ammonia Synthesis via Methane Pyrolysis

There is a growing literature and applications on the utilization of methane pyrolysis for the synthesis of common industrial chemicals such as hydrogen that is then can be used as a feedstock for ammonia synthesis. Methane pyrolysis is a process that decomposes methane (CH_4) into its elemental components hydrogen (H_2) and carbon black (C_s), without emitting carbon dioxide (CO_2) [15].

$$CH_4 \rightleftharpoons C_s + 2H_2 \quad \Delta H_{net} = 75.6 \, kJ \, mol^{-1}.$$

The implementation process of methane pyrolysis on an industrial scale requests the use of natural gas as a feedstock instead of pure methane. After the natural gas passes the pretreatment facility to remove impurities (e.g., sulfur, water, and CO_2, and even ethane, propane, and nitrogen—depending on the used catalyst), the methane-rich natural gas is fed into a bubble-column/fluidized-bed reactor where a molten metal, used as a catalyst, is readily immersed. Efforts on investigating the optimal molten metal catalyst for the pyrolysis of methane are still ongoing in which tin, bismuth-nickel, magnesium, and lead are the most reported.

At operating conditions of 800–1700 °C and 55 bar, the natural gas is fed from the bottom-up of the reactor and dispersed with the molten metal creating the required conditions for the methane to decompose. Due to the difference in density, the gas, including hydrogen, escapes the molten bath in the form of bubbles, and carbon black rises on top of the molten medium which can easily be removed, stored, and sold as a valuable by-product. The escaped gas could contain carbon particles which are captured in cyclone bags allowing the hydrogen and unreacted methane to pass into the membrane which eventually separates the product and recycles the unreacted methane.

This synthesized hydrogen could then be used for ammonia manufacture similar to the Haber-Bosch process skipping SMR process, plasma-based, or other ammonia synthesis that requires hydrogen feedstock.

4.1.6 Renewable Ammonia Synthesis

The environmental pressure exerted by the conventional Haber-Bosch process triggers scientists to look for alternative environmentally friendly approaches despite its competitively high ammonia production. The foundations for such an alternative are already in place, and renewable electricity and biomass would act as the driving force and feedstocks, respectively, instead of fossil fuels. In fact, according to IRENA 2022 Outlook [1], reports that industries are shifting towards renewable ammonia, evidently as an annual manufacturing capacity of announced ammonia plants is 15 Mt by 2030 (8% of the ammonia market across 54 projects, notably in Australia, Mauritania and Oman) [1].

Renewable ammonia, essentially, is a subsequent process of renewable hydrogen which is produced via water electrolysis. This H_2 is later on reacted with N_2 (separated from air).

The electrolysis reaction:

$$2H_2O \rightarrow 2H_2 + O_2.$$

Ammonia synthesis

$$3H_2 + N_2 \rightarrow 2NH_3.$$

Renewable ammonia is identical to ammonia produced from fossil fuels, and it is not possible to trace back its origin via any chemical analysis. Thereby, all the feedstock and energy used need to be renewable (e.g., biomass, hydro, solar, wind, geothermal) to qualify the produced ammonia to be renewable.

4.2 Ammonia Storage

Ammonia is an essential chemical that serves a crucial function in various industrial applications, such as pharmaceuticals, fertilizers, and as a potential hydrogen carrier for energy storage. Nonetheless, its chemical characteristics—particularly toxicity, corrosiveness, and high volatility—present considerable challenges for storage. Ensuring the economic viability, operational efficacy, and safety of ammonia storage systems is essential, considering these challenges. As a result, various methodologies have been developed to tackle these challenges, each providing unique benefits

4.2 Ammonia Storage

and limitations according to the particular needs of industrial processes and storage conditions.

This thorough investigation explores the principal techniques utilized for ammonia storage, providing a detailed analysis of their technical specifications and operational factors. The techniques addressed encompass solid-state storage, refrigerated storage, pressurized storage, ammonia synthesis and decomposition cycles, aqueous ammonia storage, and absorption in porous materials. Solid-state storage entails embedding ammonia within a solid matrix to diminish its volatility and improve safety, whereas refrigerated storage employs low temperatures to liquefy ammonia, consequently lowering vapor pressure and minimizing the risk of inadvertent release. Pressurized storage preserves ammonia under elevated pressure to stabilize the compound; however, this approach requires durable containment systems and incurs increased energy demands for compression. Moreover, ammonia synthesis and decomposition cycles offer dynamic regulation of ammonia production and utilization, representing a promising strategy for the incorporation of ammonia as a hydrogen carrier. Storing ammonia in solution (aqueous ammonia) provides operational advantages by mitigating the hazards linked to handling pure ammonia, while the absorption of ammonia in porous materials introduces a novel approach that utilizes advanced material characteristics to improve storage safety and efficiency.

A comprehensive understanding of these varied storage methods is crucial for enhancing the development of safer and more efficient ammonia storage systems. As the industrial applications of ammonia proliferate, additional research into the optimization of these methodologies is essential. This research seeks to address the inherent challenges posed by ammonia's chemical properties while also enhancing storage processes to facilitate its developing role in both conventional chemical industries and emerging energy technologies.

4.2.1 Pressurized Storage

Pressurized storage is one of the most widely utilized methods for storing ammonia, capitalizing on the gas's inherent ability to be maintained as a liquid under elevated pressures. This approach typically involves the use of high-pressure cylinders, spherical tanks, and bullet tanks, all of which are constructed from robust materials—such as steel or advanced composite materials—capable of withstanding pressures in the range of approximately 10–20 bar. To address the potential hazards associated with high-pressure containment, these storage vessels are outfitted with a variety of safety features, including pressure relief valves, rupture disks, and emergency shut-off systems, all of which serve to mitigate risks in the event of an overpressure situation. The inherent efficiency of pressurized storage lies in its capacity to store large quantities of ammonia in a compact volume, making it an ideal solution for industrial applications that demand immediate chemical availability while operating under spatial constraints. Nonetheless, the high-pressure environment necessitates meticulous maintenance and continuous monitoring to prevent mechanical failures, such as bursts or leaks, which could otherwise result in significant safety hazards.

4.2.2 Refrigerated Storage

Refrigerated storage is a vital technological method for reducing the risks linked to high-pressure ammonia storage. This technique guarantees that ammonia stays in a liquid form at atmospheric pressure by keeping it at temperatures below its boiling point of $-33.34\,°C$. To accomplish this, extensive insulated storage tanks are utilized, typically featuring double-walled constructions and supplementary insulation layers, such as vacuum or inert gas barriers, to reduce thermal exchange with the external environment. These refrigerated tanks are additionally outfitted with sophisticated temperature monitoring systems and pressure relief mechanisms to guarantee secure and stable storage conditions.

The principal benefit of refrigerated storage is its improved safety characteristics. Maintaining ammonia at low pressures substantially diminishes the risk of catastrophic failures, including tank ruptures or leaks. This is especially critical due to the dangerous characteristics of ammonia, which presents significant threats to human health and the environment in the case of a release. This method presents certain challenges. The necessity for prolonged low temperatures demands a constant energy supply, rendering the process fundamentally energy-intensive. The advanced insulation systems necessary to reduce heat ingress require regular maintenance and operational supervision, thereby elevating the complexity and expense of this storage method.

In conclusion, refrigerated storage provides a safer option than high-pressure storage for ammonia by keeping it in a liquid form at atmospheric pressure; however, it entails considerable energy and maintenance requirements. These factors must be meticulously assessed when determining the viability and sustainability of this storage method in industrial applications.

4.2.3 Solid-State Storage

Solid-state storage constitutes a novel method for ammonia storage, utilizing materials that can chemically bind or absorb ammonia in a stable, solid state. This method employs various materials, such as metal halides, metal amides, and complex hydrides, each possessing distinct properties that facilitate effective ammonia absorption. Materials like magnesium chloride ($MgCl_2$) and lithium amide ($LiNH_2$) exhibit exceptional ammonia absorption capabilities, creating stable compounds that can later release ammonia under regulated conditions. The reversible absorption-desorption mechanism renders solid-state storage a viable alternative to traditional liquid or high-pressure storage methods.

A primary benefit of solid-state storage is its superior safety characteristics. By encapsulating ammonia within a solid matrix, the hazards related to leaks, spills, or gaseous emissions are significantly reduced. This enhances operational safety and facilitates the handling and transportation of ammonia, rendering it a desirable choice for diverse industrial applications. Moreover, the solid-state characteristic of this storage technique obviates the necessity for high-pressure containment or cryogenic conditions, thereby diminishing energy demands and infrastructural intricacy.

Notwithstanding these benefits, solid-state storage technologies are still in the developmental phase and encounter numerous technical obstacles. Critical concerns encompass capacity diminution following successive absorption-desorption cycles and material deterioration resulting from chemical instability or mechanical strain. These constraints presently impede the extensive adoption of solid-state storage in commercial applications. As a result, current research endeavors aim to enhance the stability, adsorption capacity, and cyclability of these materials. Progress in materials science, including the creation of innovative composites and nanostructured materials, offers potential for addressing these challenges and positioning solid-state storage as a feasible and sustainable method for ammonia storage in the future.

In conclusion, although solid-state storage presents a secure and efficient option for ammonia storage, additional improvements in material performance and durability are necessary to achieve its full potential. Ongoing research and innovation in this domain are essential for overcoming current limitations and facilitating the commercialization of this promising technology.

4.2.4 Liquid Ammonia

The storage of ammonia in solution is possible as liquid ammonia or ammonium hydroxide (NH_4OH), which entails the dissolution of ammonia gas (NH_3) in water (H_2O). This method is especially beneficial for the storage and management of limited amounts of ammonia, as it alleviates numerous risks linked to pure ammonia gas, including high volatility and toxicity. Liquid ammonia is generally stored in corrosion-resistant containers, commonly made from materials like high-density polyethylene (HDPE) or stainless steel, chosen for their durability and chemical resistance. The ammonia concentration in these solutions typically varies from 10 to 30% by weight, ensuring a compromise between storage safety and functional applicability. This storage method offers a superior safety profile. Dissolving ammonia in water results in a solution with markedly diminished volatility relative to pure liquid ammonia, facilitating handling and transportation. Furthermore, the aqueous form reduces the likelihood of inadvertent emissions of gaseous ammonia, which can present significant health and environmental risks. Nonetheless, this methodology also exhibits specific constraints. The reduced concentration of ammonia in solution, relative to pure liquid ammonia, requires supplementary processing stages to extract and concentrate ammonia for industrial or commercial applications. This may elevate both the intricacy and expense of the storage system.

Containers for liquid ammonia must be fitted with suitable venting systems to guarantee safe and effective storage. These systems are essential for averting pressure buildup within the storage vessel, which may occur from the gradual emission of ammonia gas due to temperature variations or shifts in chemical equilibrium. Effective venting systems ensure stable storage conditions and facilitate the regulated release of ammonia gas as necessary.

In conclusion, the storage of ammonia in liquid solution provides a safer and more controllable option for managing smaller amounts of ammonia, especially in scenarios where elevated concentrations are unnecessary. This method mitigates the risks linked to pure ammonia gas, yet it also presents challenges concerning ammonia concentration and extraction. The design and implementation of resilient storage systems, incorporating corrosion-resistant materials and efficient venting mechanisms, are crucial for ensuring the safety and efficacy of this storage method.

4.2.5 Absorption in Porous Materials

The employment of high-surface-area materials, including activated carbon, metal-organic frameworks (MOFs), zeolites, silica gel, and carbon nanotubes, signifies a novel and developing method for ammonia storage via adsorption in porous matrices. These materials demonstrate remarkable potential for ammonia storage owing to their extensive pore structures, elevated surface areas, and adjustable physicochemical properties, facilitating the adsorption of substantial amounts of ammonia. MOFs and activated carbon are notably efficient for ammonia adsorption due to their remarkably high surface areas and the capacity to customize their pore sizes and surface chemistries to enhance adsorption efficacy. This method provides significant benefits regarding safety and stability, as the porous matrix effectively contains ammonia within its framework, thereby markedly diminishing the likelihood of leaks or accidental discharges.

This approach offers a superior safety profile compared to traditional storage methods, including high-pressure or cryogenic storage. By encapsulating ammonia within a solid adsorbent, the hazards linked to gaseous or liquid ammonia, such as volatility and toxicity, are significantly reduced. The solid-state characteristics of these materials facilitate handling and transportation, rendering them appropriate for various applications, including industrial processes and energy storage systems. Nonetheless, the creation and execution of these materials remain in the research and optimization stage. Challenges, including the optimization of adsorption capacities, the maintenance of material stability through multiple adsorption-desorption cycles, and the attainment of cost-effective scalability, persist as significant obstacles to their extensive implementation.

Current research endeavors aim to tackle these challenges by developing advanced materials that exhibit superior adsorption kinetics, increased ammonia storage capacities, and enhanced durability. Advancements in material design, including the integration of functional groups and the creation of hybrid composites, are being investigated to improve the efficacy of these porous materials. Moreover, incorporating these materials into functional storage systems necessitates meticulous evaluation of aspects such as regeneration efficiency, energy demands, and long-term material durability.

In conclusion, the utilization of high-surface-area porous materials for ammonia storage presents considerable potential for transforming the storage and management of ammonia in diverse applications. Despite ongoing challenges concerning material performance and scalability, further progress in material science and engineering is anticipated to propel the commercialization of this technology. By overcoming these limitations, porous materials can offer a safer, more efficient, and sustainable solution for ammonia storage, thereby advancing energy storage and industrial processes. Figure 4.4 shows ammonia (NH_3) storage techniques into five primary types: pressurized storage, refrigerated storage, solid-state storage, liquid ammonia storage, and absorption in porous materials. Pressurized storage includes high-pressure cylinders, spherical tanks, and bullet tanks, while refrigerated storage utilizes refrigerated and double-walled tanks for maintaining ammonia in a liquid state. Solid-state storage relies on metal halides, metal amides, and complex hydrides such as magnesium chloride, calcium chloride, and lithium aluminum hydride for stable storage. Liquid ammonia can be stored in tanks, drums, and intermediate bulk containers, while advanced porous materials like activated carbon, zeolites, silica gel, and carbon nanotubes provide efficient absorption-based storage solutions.

4.3 Ammonia Transport

Efficient and secure transportation methods are necessary for ammonia, which is a crucial industrial chemical and a potential hydrogen carrier, because of its extensive use and hazardous properties. Ammonia is commonly transported as a liquid without water, and it is kept at a temperature of − 33 °C under moderate pressure. Ammonia transportation, like hydrogen transportation, can be classified according to its physical form, either as a gas or a liquid. Ammonia transportation methods are as follows.

4.3.1 Pressurized Tankers and Barges

Liquid ammonia is an essential industrial chemical commonly transported over long distances by maritime means utilizing specialized pressurized tankers and barges. These vessels are precisely designed to securely transport substantial quantities of

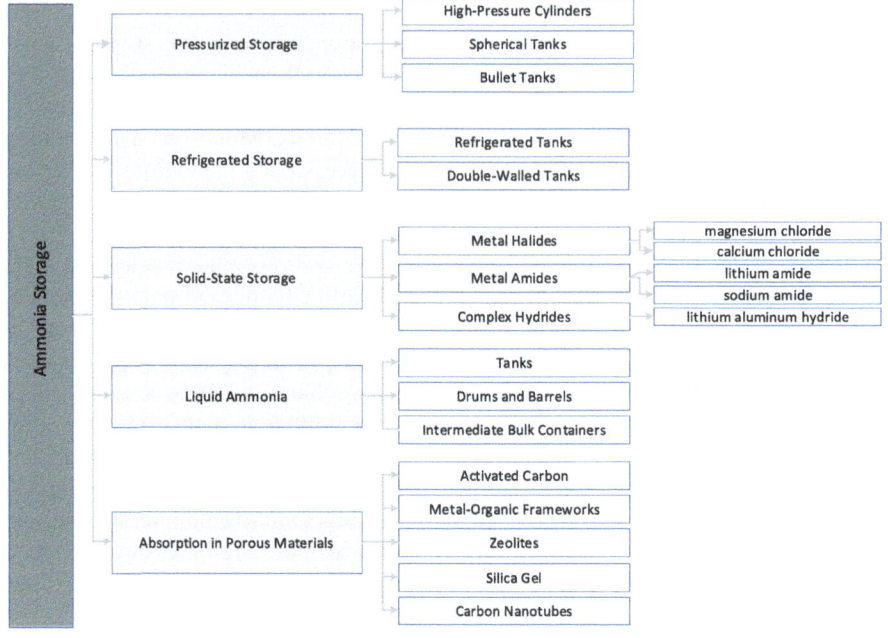

Fig. 4.4 Ammonia storage methods and classification

ammonia, with tankers accommodating up to 20,000 metric tons and barges generally managing smaller loads between 500 and 5000 metric tons [16]. The cargo tanks on these vessels feature sophisticated refrigeration and insulation systems to preserve ammonia in its liquid form at approximately − 33 °C, its boiling point under atmospheric pressure. This guarantees the stability of ammonia and reduces the likelihood of vaporization during transportation.

Safety is the foremost priority in the maritime transportation of ammonia, and contemporary vessels are engineered with multiple protective layers to reduce potential risks. This encompasses advanced gas detection systems for leak monitoring, emergency venting mechanisms to avert over-pressurization, and resilient pressure control systems to ensure optimal storage conditions. Furthermore, the design and operation of these vessels are regulated by rigorous international standards, particularly the International Maritime Organization's (IMO) International Code for the Construction and Equipment of Ships Carrying Liquefied Gases in Bulk (IGC Code). The IGC Code sets forth extensive safety standards for the construction, equipment, and operation of gas carriers, guaranteeing the safe and efficient transport of liquefied gases, such as ammonia, over oceans and inland waterways [17].

The IGC Code requires stringent inspection and certification procedures for vessels, along with the establishment of emergency response protocols to manage potential incidents during transit. These measures are essential for protecting human health and the environment due to the hazardous characteristics of ammonia, which is toxic, corrosive, and potentially explosive under specific conditions. Moreover, the code underscores the significance of crew training and the implementation of advanced technologies to improve operational safety and environmental safeguarding.

Besides safety considerations, the maritime transport of ammonia is also affected by economic and logistical factors. The capacity to transport substantial volumes of ammonia efficiently across extensive distances renders maritime shipping a financially advantageous choice for international trade. The energy-intensive requirements for sustaining cryogenic conditions and the necessity for specialized infrastructure, including port facilities capable of managing liquefied gases, pose persistent challenges. Innovations in vessel design, including the creation of energy-efficient refrigeration systems and the utilization of lightweight, corrosion-resistant materials, are mitigating these challenges and enhancing the sustainability of ammonia transportation.

In summary, the maritime transportation of liquid ammonia is a multifaceted and stringently regulated procedure that emphasizes safety, efficiency, and environmental safeguarding. The utilization of specialized tankers and barges, along with sophisticated safety systems and compliance with international regulations like the IGC Code, guarantees the safe and dependable transportation of ammonia to worldwide markets. Ongoing advancements in vessel design and operational methodologies will be crucial for satisfying the increasing demand for ammonia while mitigating risks and environmental consequences.

4.3.2 Pressurized Rail Tank Cars and Tank Trucks

Ammonia land transportation is primarily conducted via pressurized rail tank cars and tank trucks, which provide notable benefits regarding volume capacity, flexibility, and access to diverse delivery sites. Rail tank cars, typically capable of holding up to 30,000 gallons (113,562 l), are especially adept at the efficient transportation of substantial volumes of ammonia across extended distances. This transportation method is particularly beneficial in areas with developed rail systems, as it offers a cost-efficient and dependable means of delivering ammonia to industrial and agricultural centers. The substantial capacity of rail tank cars decreases the number of trips needed, thus lowering transportation expenses and environmental effects related to fuel usage and emissions [18].

Conversely, tank trucks, typically possessing a lower capacity than rail tank cars, provide enhanced adaptability and maneuverability, rendering them optimal for ammonia delivery to designated sites, especially in regions lacking pipeline infrastructure. Tank trucks are frequently employed for last-mile delivery, guaranteeing that ammonia arrives at end-users, including fertilizer manufacturers, refrigeration facilities, and chemical plants. Their capacity to traverse local road networks and reach remote or rural regions renders them an essential element of the ammonia supply chain.

Rail tank cars and tank trucks are engineered with safety as a paramount consideration. Their tanks are fabricated from high-strength materials and incorporate advanced safety features, such as pressure relief valves and thermal insulation systems. These characteristics guarantee that ammonia retains a stable liquid form during transportation, regardless of fluctuating environmental conditions. Pressure relief valves are essential for averting over-pressurization, which may result in tank ruptures or leaks, whereas thermal insulation is necessary to sustain the requisite low temperatures to inhibit ammonia vaporization. Moreover, the tanks are frequently constructed with double walls or reinforced structures to offer an additional safeguard against possible accidents or impacts.

The conveyance of ammonia via rail and road is governed by rigorous regulatory scrutiny to guarantee safety and environmental safeguarding. The U.S. Department of Transportation (DOT) in the USA enforces extensive regulations regarding the design, construction, and operation of ammonia transport vehicles [19]. The regulations are specified in the Code of Federal Regulations (CFR), especially Title 49, which pertains to the transportation of hazardous materials. In Europe, the European Agreement concerning the International Carriage of Dangerous Goods by Road (ADR) sets standardized safety protocols for the road transport of ammonia and other hazardous materials [20]. The ADR delineates stipulations for vehicle design, labeling, documentation, and driver training, guaranteeing the safe transportation of ammonia across international boundaries.

Notwithstanding the stringent safety protocols implemented, the terrestrial conveyance of ammonia presents certain challenges. The perilous characteristics of ammonia require meticulous management and oversight during transportation. Although infrequent, accidents can result in significant repercussions, including environmental pollution and threats to public health. Consequently, continuous endeavors are directed toward enhancing safety technologies, including the advancement of sophisticated pressure relief systems, superior tank materials, and real-time monitoring systems to identify leaks or pressure fluctuations during transportation.

In summary, the terrestrial transport of ammonia through rail tank cars and tank trucks is an essential element of the global ammonia supply chain, providing a combination of efficiency, adaptability, and safety. Compliance with stringent regulatory frameworks, such as those set forth by the U.S. DOT and the ADR, guarantees the safe and reliable transportation of ammonia. Ongoing progress in vehicle design and safety technologies will be crucial for tackling the challenges related to ammonia transportation and preserving the integrity of this essential industrial process.

4.3.3 Pipelines

Ammonia transportation through pipelines is less frequent compared to other methods, mainly because ammonia is corrosive. However, there are specific pipelines designed for the transportation of ammonia, especially in areas where production facilities are directly linked to major consumption centers. The pipelines are made of durable materials, such as carbon steel, and are specifically designed to meet the needs of transporting ammonia over long distances. Pipelines provide a cost-efficient and effective method for the continuous and high-volume transportation of ammonia. Nevertheless, the utilization of these systems is constrained by the substantial upfront capital needed and possible ecological issues associated with land disturbance. Regular surveillance and upkeep are crucial to guarantee the secure functioning of ammonia pipelines.

Alternative modes of transportation are transporting ammonia as a hydrogen carrier. Ammonia is recognized as a promising hydrogen carrier, in addition to its primary use in industries. Ammonia is more energy-dense and requires less stringent storage conditions compared to liquid hydrogen, making it a favorable choice for transporting hydrogen. However, the use of ammonia as a hydrogen carrier necessitates the establishment of additional infrastructure to perform the process of cracking ammonia back into hydrogen at the location where it is needed. This process entails the introduction of intricacy and supplementary expenses, along with the possibility of energy dissipation during the conversion. Ammonia's potential as a hydrogen carrier is currently being actively investigated, especially for large-scale hydrogen transport and storage solutions, despite the challenges involved. Table 4.2 provides a comparative overview of the different transportation modes for ammonia, highlighting their advantages and disadvantages. Additionally, the choice of transportation method for ammonia is contingent upon several factors, such as the distance to be traveled, the quantity of ammonia to be conveyed, the presence of necessary infrastructure, and considerations pertaining to cost, safety, and environmental consequences. Pressurized tankers are optimal for transporting large quantities over long distances by sea, whereas tank trucks are more suitable for accessing isolated or specific locations. Pipelines, when accessible, provide a highly efficient solution for the continuous and large-scale transportation of ammonia. However, they necessitate substantial upfront investment and ongoing maintenance.

Advancements in transportation technology will be crucial as the demand for ammonia increases, driven by its traditional industrial uses and its emerging role as a hydrogen carrier. Continuous advancements in tank design, safety systems, and logistical strategies are anticipated to improve the safety and effectiveness of ammonia transportation. In addition, the establishment of infrastructure to facilitate the utilization of ammonia as a means of transporting hydrogen could have a substantial impact on the future energy sector, presenting novel prospects for sustainable, large-scale energy transportation solutions.

Table 4.2 Overview of transportation modes for ammonia: advantages and disadvantages

Mode of transportation	Advantages	Disadvantages
Maritime (tankers)	High capacity, suitable for long distances, relatively low cost per unit	High initial investment, potential for environmental damage in case of spills, limited accessibility to inland locations
Maritime (barges)	Cost-effective for inland waterways, flexible	Limited capacity, restricted to navigable waterways, slower than rail or pipeline
Land (rail tank cars)	Efficient for bulk transport over land, well-established infrastructure, relatively safe	Limited flexibility, potential for accidents and spills, requires access to rail network
Land (tank trucks)	Versatile, suitable for localized deliveries, flexible routing	Lower capacity, higher cost per unit compared to rail or pipeline
Land (pipelines)	Cost-effective for long distances, minimal environmental impact during operation, high capacity	High initial investment, potential for leaks and land disruption, limited existing infrastructure
Ammonia as hydrogen carrier	Higher energy density than liquid hydrogen, less demanding storage conditions	Requires additional infrastructure for hydrogen extraction, energy losses during conversion

References

1. IEA (2022) Innovation outlook: renewable ammonia. Int Renew Energy Agency 129
2. Al-Breiki M, Bicer Y (2021) Comparative life cycle assessment of sustainable energy carriers including production, storage, overseas transport and utilization. J Clean Prod 279. https://doi.org/10.1016/j.jclepro.2020.123481
3. Aziz M, TriWijayanta A, Nandiyanto ABD (2020) Ammonia as effective hydrogen storage: a review on production, storage and utilization. Energies 13. https://doi.org/10.3390/en13123062
4. Jiao F, Xu B (2019) Electrochemical ammonia synthesis and ammonia fuel cells. Adv Mater 31. https://doi.org/10.1002/adma.201805173
5. Lan R, Irvine JTS, Tao S (2013) Synthesis of ammonia directly from air and water at ambient temperature and pressure. Sci Rep 3. https://doi.org/10.1038/srep01145
6. Murakami T, Nishikiori T, Nohira T, Ito Y (2003) Electrolytic synthesis of ammonia in molten salts under atmospheric pressure. ChemInform 34. https://doi.org/10.1002/chin.200318020
7. Knani S, Holade Y, Basile A (2024) Electrochemical synthesis of ammonia. Prog Ammon Sci Technol Membr 63–88. https://doi.org/10.1016/B978-0-323-88516-4.00006-8
8. Marnellos G, Zisekas S, Stoukides M (2000) Synthesis of ammonia at atmospheric pressure with the use of solid state proton conductors. J Catal 193:80–87. https://doi.org/10.1006/jcat.2000.2877
9. Skodra A, Ouzounidou M, Stoukides M (2006) NH3 decomposition in a single-chamber proton conducting cell. Solid State Ionics 177:2217–2220. https://doi.org/10.1016/j.ssi.2006.03.051
10. Liu RQ, Xie YH, Wang J De, Li ZJ, Wang BH (2006) Synthesis of ammonia at atmospheric pressure with $Ce_{0.8}M_{0.2}O_{2-\delta}$ (M = La, Y, Gd, Sm) and their proton conduction at intermediate temperature. Solid State Ionics 177:73–76. https://doi.org/10.1016/j.ssi.2005.07.018

References

11. Amar IA, Lan R, Petit CTG, Arrighi V, Tao S (2011) Electrochemical synthesis of ammonia based on a carbonate-oxide composite electrolyte. Solid State Ionics 182:133–138. https://doi.org/10.1016/j.ssi.2010.11.009
12. Hollevoet L, Jardali F, Gorbanev Y, Creel J, Bogaerts A, Martens JA (2020) Towards green ammonia synthesis through plasma-driven nitrogen oxidation and catalytic reduction. Angew Chemie 132:24033–24037. https://doi.org/10.1002/ange.202011676
13. Hollevoet L, Jardali F, Gorbanev Y, Creel J, Bogaerts A, Martens JA (2020) Towards green ammonia synthesis through plasma-driven nitrogen oxidation and catalytic reduction. Angew Chemie Int Ed 59:23825–23829. https://doi.org/10.1002/anie.202011676
14. Rouwenhorst KHR, Jardali F, Bogaerts A, Lefferts L (2021) From the Birkeland-Eyde process towards energy-efficient plasma-based NO_X synthesis: a techno-economic analysis. Energy Environ Sci 14:2520–2534. https://doi.org/10.1039/D0EE03763J
15. Parkinson B, Matthews JW, McConnaughy TB, Upham DC, McFarland EW (2017) Techno-economic analysis of methane pyrolysis in molten metals: decarbonizing natural gas. Chem Eng Technol 40:1022–1030. https://doi.org/10.1002/ceat.201600414
16. GIZ (2024) Ammonia transport and storage. Dtsch Gesellschaft Für Int Zusammenarbeit GmbH 1:1–13
17. IMO (2012) International code for the construction and equipment of ships carrying liquefied gases in bulk (IGC Code). https://www.imo.org/en/ourwork/environment/pages/igccode.aspx
18. Nayak-Luke RM, Forbes C, Cesaro Z, Bánares-Alcántara R (2020) Techno-economic aspects of production, storage and distribution of ammonia. In: Techno-economic challenges green ammonia as an energy vector, pp 191–207. https://doi.org/10.1016/B978-0-12-820560-0.00008-4
19. DOT (2023) Code of federal regulations, Title 49: transportation of hazardous materials. Washington, D.C
20. UNECE (2018) European agreement concerning the international carriage of dangerous goods by road. Geneva. https://doi.org/10.18356/903cdddb-en

Open Access This chapter is licensed under the terms of the Creative Commons Attribution 4.0 International License (http://creativecommons.org/licenses/by/4.0/), which permits use, sharing, adaptation, distribution and reproduction in any medium or format, as long as you give appropriate credit to the original author(s) and the source, provide a link to the Creative Commons license and indicate if changes were made.

The images or other third party material in this chapter are included in the chapter's Creative Commons license, unless indicated otherwise in a credit line to the material. If material is not included in the chapter's Creative Commons license and your intended use is not permitted by statutory regulation or exceeds the permitted use, you will need to obtain permission directly from the copyright holder.

Chapter 5
Methanol

5.1 Methanol Production

Methanol (CH_3OH) is one of the most critical chemicals—alongside ethylene, propylene and ammonia—used to synthesize other compounds in the chemical industry. Formaldehyde, acetic acid, and plastics/polymers are common methanol-based products. Figure 5.1 illustrates various methanol synthesis pathways, categorized into non-renewable sources (natural gas, oil, and coal) and renewable sources (water, sunlight, heat, and biomass). Non-renewable methods include steam methane reforming, methane pyrolysis, partial oxidation, plasma reforming, and gasification, while renewable methods utilize electrolysis, photobiological, photoelectrochemical, thermochemical, and biomass gasification processes. The production of syngas (a mixture of hydrogen and carbon monoxide) or direct hydrogen (H_2) and carbon dioxide (CO_2) from these sources enables methanol synthesis, which plays a crucial role in the chemical industry, fuel applications, and energy storage.

5.1.1 Steam Methane Reforming

Shaping 65% of the global production, the natural gas route is at the forefront of methanol synthesis due to its material availability and relatively straightforward manufacturing process. The process involves 3 main steps: production of synthesis gas, conversion of syngas into crude methanol, distillation of the reactor effluent to achieve the desired purity [1]. Figure 5.2 outlines the methanol production process using natural gas as the primary feedstock. The process begins with feed preparation and desulfurization to remove impurities. The steam reforming step converts natural gas into synthesis gas (syngas), a mixture of hydrogen (H_2), carbon monoxide (CO), and carbon dioxide (CO_2). The syngas then undergoes methanol synthesis, producing crude methanol, which is further refined through methanol distillation to yield pure

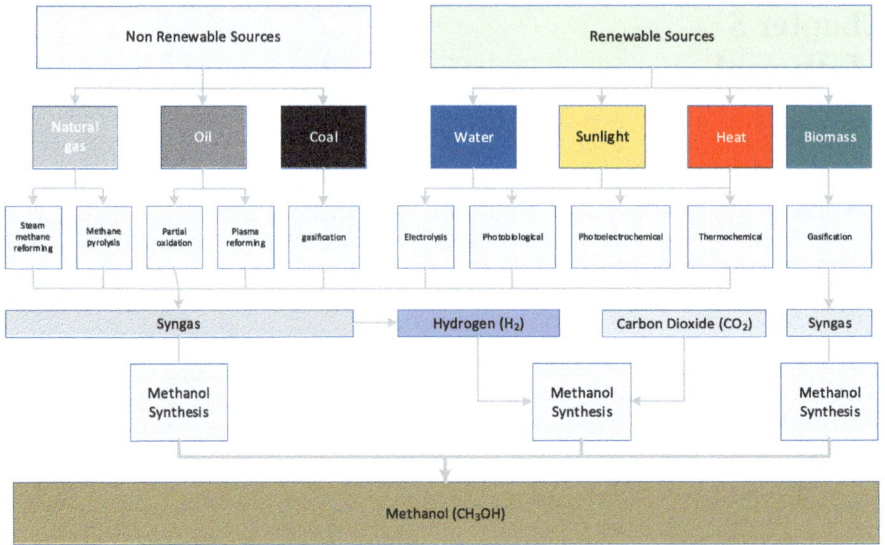

Fig. 5.1 Methanol production pathways from non-renewable and renewable sources

methanol (CH_3OH), a valuable chemical used in fuels, solvents, and various industrial applications.

Production of synthesis gas

The mixture of synthesis gas consisting of (H_2, CO, and CO_2) is mainly produced by steam methane reforming (SMR) as well as autothermal reforming (ATR) of natural gas.

$$CH_4 + H_2O \rightarrow CO + 3H_2$$
$$CH_4 + 2O_2 \rightarrow CO_2 + 2H_2O$$

The production of both hydrogen and carbon dioxide leads to a water gas shift (WGS) reaction and partial oxidation due to the presence of oxygen.

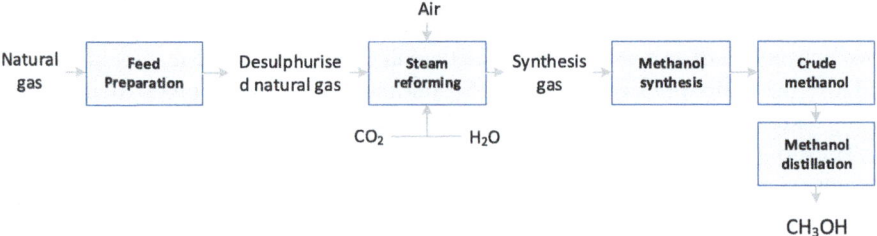

Fig. 5.2 Methanol production process from natural gas

5.1 Methanol Production

$$CO_2 + H_2 \rightleftharpoons CO + H_2O$$

$$CH_4 + \frac{1}{2}O_2 \rightarrow CO + 2H_2$$

Under ideal conditions, the production of methanol should have a syngas composition (S); the difference between H_2 and CO_2 moles, and the summation of the moles of CO_2 and CO, ratio of 2.

$$S = \frac{\text{moles } H_2 - \text{moles } CO_2}{\text{moles CO} + \text{moles } CO_2}$$

The stoichiometric value S takes into account the presence of CO_2 converted that consumes H_2 via reverse WGS reaction.

$$CO_2 + 3H_2 \rightleftharpoons CH_3OH + H_2O, \quad \Delta H = 41.17 \frac{kJ}{mol}$$

However, the value of S is raw material-dependent. When a syngas is produced by SMR, an S value of 2.8–3 is usually obtained.

Conversion of syngas into crude methanol

According to [2], the process converting syngas into crude methanol takes place at a temperature and pressure ranges of 200–300 °C and 50–100 bar, respectively. Hydrogenation of CO and CO_2 are the main reactions.

$$CO + 2H_2 \rightleftharpoons CH_3OH, \quad \Delta H = -100.46 \frac{kJ}{mol}$$

which is divided into 2 steps

$$CO + H_2 \rightleftharpoons CH_2O$$
$$CH_2O + H_2 \rightleftharpoons CH_3OH$$

$$CO_2 + 3H_2 \rightleftharpoons CH_3OH + H_2O, \quad \Delta H = -61.59 \frac{kJ}{mol}.$$

The mixed gases are then fed to the converter including H_2/CO, maintained at ratio from 3:1 to 5:1 for the conventional gas-phase process which often requires equipment where the WGS is used to accelerate the hydrogen content. The liquid phase methanol process can handle the synthesis gas straight from the generator thanks to its superior heat management capabilities in 1:1 to 1:2 ratios. As indicated by the negative enthalpy signs, however, both hydrogenation reactions are highly exothermic, thereby, require significant cooling. This thermodynamic equilibrium leads to low conversion per pass, implying a large recycle of unconverted gas which are responsible for the investment costs of this process segment. Finally, the cooled gas mixture is conveyed into the distillation columns to produce pure methanol.

5.1.2 Coal Gasification

About a half of the global methanol production capacity and consumption resides in China where most of which produced via coal to methanol CTM route. The CTM route adopts coal gasification process in order to produce the syngas (CO and H_2 mixture). There are two different types of reactions to coal gasification occurs: heterogenous reactions at the phase interface, in which the gasifying agent and the product gases react with the coal, and homogenous reactions, in which the primary oxidation gases, and the devolatilization gases released by the coal react with each other [3].

Heterogenous reactions (gas/solids)

$$C_s + H_2O \rightarrow CO + H_2, \quad \Delta H = 119 \frac{kJ}{mol}$$

$$C_s + CO_2 \rightarrow 2CO, \quad \Delta H = 162 \frac{kJ}{mol}$$

$$C_s + 2H_2 \rightarrow CH_4, \quad \Delta H = -87 \frac{kJ}{mol}$$

$$C_s + \frac{1}{2}O_2 \rightarrow CO, \quad \Delta H = -123 \frac{kJ}{mol}$$

$$C_s + O_2 \rightarrow CO_2, \quad \Delta H = -406 \frac{kJ}{mol}$$

Homogenous reactions (gas/gas)

$$CO + H_2O \rightarrow CO_2 + H_2, \quad \Delta H = -42 \frac{kJ}{mol}$$

$$CO + 3H_2 \rightarrow CH_4 + H_2O, \quad \Delta H = -206 \frac{kJ}{mol}.$$

It is the interplay of these reactions that determines the composition of the product gas. Depending on the type of gasifier used, these reactions may occur at clearly different times and locations (fixed-bed gasifier) or almost simultaneously (high-temperature dust gasification). For methanol production, the diverse utilization of syngas is highly favorable thereby, the discussed homogenous reactions are often limited in the CTM plant designs by focusing on Boudouard and steam reactions [4]. This is achieved by injecting steam (H_2O) and carbon dioxide (CO_2) (i.e., the initial concentration of CO_2 pushes the homogenous reactions backwards and vice versa) with the char content (C_s) into the gasifier at operating conditions of 900 °C and 20 bar at least. The high operating pressure has several benefits; (1) it favors the char conversion into syngas according to Le Chaterlier's Principle and (2) the compression duties of produced syngas to enter the fuel synthesis reactor are minimized when compared to gasification at atmospheric pressure. As the desired reactions are endothermic, the required heat for the gasification is supplied from a char combustor

5.1 Methanol Production

by heat pipes. The temperature difference between the two reactors determines the ratio of the char used for gasification, the char that is used for combustion.

$$CO + 2H_2 \rightleftharpoons CH_3OH, \quad \Delta H = -100.46 \frac{kJ}{mol}$$

$$CO_2 + 3H_2 \rightleftharpoons CH_3OH + H_2O, \quad \Delta H = -61.59 \frac{kJ}{mol}$$

Lastly, hydrogenation of carbon monoxide (CO) and carbon dioxide takes place in the methane reactor to convert the syngas into methanol and water which is ultimately separated by distillation columns.

5.1.3 Biomass Gasification

Methanol produced from biomass is a promising carbon-neutral fuel. In fact, it is well-suited for use in fuel cell vehicles (FCVs), which are anticipated to operate at significantly higher efficiency levels compared to current Internal Combustion Engine Vehicles (ICEVs). Additionally, bio-methanol is considered relatively clean, emitting none of the air pollutants SO_x, NO_x, VOS or dust when fully combusted apart from CO_2. Nonetheless, it is derived from sustainably grown biomass, the overall energy chain can be GHG neutral. Such fuel alternative could play a major role in the transportation sector as the globe transitions away from non-renewables [5]. Figure 5.3 illustrates the methanol production process using biomass as a renewable feedstock. The process begins with biogas production, where biomass is converted into biogas through anaerobic digestion. The biogas undergoes pretreatment to remove impurities, producing carbon dioxide (CO_2) and purified gas. A steam reformer converts the treated biogas into syngas (a mixture of H_2, CO, and CO_2), which is then used in methanol synthesis to produce crude methanol. Finally, methanol distillation refines the product, yielding pure methanol (CH_3OH) for use as a fuel, chemical feedstock, and energy carrier.

The industrial production of bio-methanol typically consists of basic processes: Pretreatment, gasification, gas cleaning, reforming of higher hydrocarbons, shift

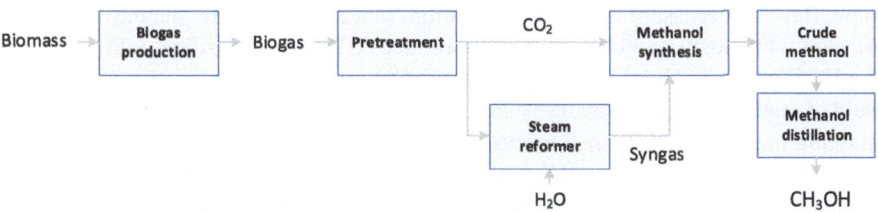

Fig. 5.3 Methanol production from biomass via biogas pathway

to obtain appropriate H_2:CO ratios, and gas separation for methanol synthesis and purification.

Pretreatment

There is a vast range of biomass materials that could be used for bio-methanol production. Depending on the type of biomass feedstock, the complexity of the pretreatment facility is determined. Generally, chipping or comminution is the first step in biomass preparation—unless pure biogas is used, chipping is neglected. The fuel size necessary for fluidized bed gasification is between 0 and 50 mm [6]. For one kg of wet biomass, approximately 100 kJ energy required to chip down woody biomass to 3×3 cm in a hammermill. Next, the fuel is usually dried to 10–15% depending on the type of gasifier. To conserve energy and cut out costs, drying by the hot flue gas or steam is often suggested, a choice that, among others, depends on other steam demands within the process and the extent of electricity co-production. Although it consumes 10% of the energy content, flue gas drying gives bio-methanol production higher flexibility towards gasification of a large variety of fuels. However, flue gas drying holds a risk of spontaneous combustion and corrosion. Thus, often plant designs settle on the utilization of an elaborate steam cycle for an inherently safe design alternative to flue gas drying. The dried biomass is then fed into the gasifier.

Gasification

The purpose of the gasifier is to convert solid biomass into syngas similar to the above discussed methanol processes [7]. A mixture of CO, CO_2, H_2O, H_2, and light hydrocarbons, syngas is produced in which the mutual ratios depend on the type of biomass, gasifier type, temperature and pressure, and the use of air, oxygen, and steam.

Many gasification methods are available for synthesis gas production. Based on throughput, cost, complexity, and efficiency issues, circulated fluidized bed gasifiers are the most suitable for large-scale synthesis gas production.

The Institute of Gas Technology (IGT) and the Battle Columbus (BCL) gasifiers, both are highly suitable for industrial scale production of syngas. While both aim to produce syngas, they differ in the method of heating biomass via direct and indirect heating, respectively.

The IGT gasifier is directly heated by burning some of the biomass containment to provide the necessary heat for gasification. In a fluidized state, the gasifier bed is injected with steam and oxygen which produces a fluent gas stream with an attractive H_2:CO ratio of (1.4:1) although the large CO_2 concentration lowering the overall yield of methanol. This is favorable due to the generated large pressure in the gasifier enabling large throughput per reactor volume and diminishes the need for pressurization downstream. Additionally, the bottom-up injection of oxidants enables high mixing levels for the gasification to occur over the whole bed at a relatively uniform temperature of (800–1000 °C). Ash, unreacted char and particulates are entrained within the product gas and largely removed using a cyclone. The product gas exits essentially at bed temperature and conveyed to the next stage process. An important

5.1 Methanol Production

characteristic of the IGT synthesis gas is the relatively large CO_2 and CH_4 fractions which are results of direct heating and the non-equilibrium nature of biomass gasification and pressurized operations, respectively.

BCL indirectly heats the gasifier containment by a heat transfer mechanism. Similarly, ash, char and sand are entrained in the product gas and separated using a cyclone, but sent to a second bed where the char or additional biomass is burned in air to reheat the sand. This enables the heat transfer between the two vessels where one facilitates combustion heat and the other gasification by circulating the hot sand back to the gasification bed without the need to use oxygen. Operating at atmospheric pressure, BCL gasifier produces a gas with leaner CO_2 and richer hydrocarbon concentrations. Therefore, tar cracking and reforming are logical subsequent processes in order to maximize CO and H_2 production. The reactor is swiftly fluidized allowing throughputs equal to the bubbling fluidized IGT albeit the atmospheric operation. Thanks to these operating conditions, the associated costs: scaling, maintenance, and others, are minimal as more BCL commercial experiences are gained. On the other hand, significant departures from the equilibrium are often found in the product gas due to the relatively low gasification temperatures.

Gas Cleaning

Using available conventional technology, by applying gas cooling, low temperature filtration and water scrubbing at 100–250 °C, the gas can be cleaned from residual impurities that could poison the catalyst in the reactor. Another alternative is hot gas cleaning using ceramic filters and reagents at 350–800°C can be considered. It is important to highlight that the considered pressure range is not a problem for either technologies. For innovative heat integration, however, hot gas cleaning could be advantageous when a reformer is applied directly after the cleaning section albeit it is not yet proven to be effective.

Reforming and WGS

The syngas can contain considerable amount of methane (CH_4) and light hydrocarbons, as abovementioned, representing a significant part of the heating value of the gas. Driven by steam addition over typically a nickel-based catalyst, SMR converts these compounds to CO and H_2 at high temperatures [7].

$$CH_4 + H_2O \rightarrow CO + 3H_2$$
$$C_2H_4 + 2H_2O \rightarrow 2CO + 4H_2$$
$$C_2H_6 + 2H_2O \rightarrow 2CO + 5H_2$$

Concurrently, water gas shift (WGS) takes place and brings the reformer into chemical equilibrium.

$$CO + H_2O \rightarrow CO_2 + H_2$$

Although reforming is favored at lower pressures, elevated pressures are economically beneficial in small equipment. At temperature and pressure ranges of 830–1000 °C and 1–3.5 MPa, respectively, reformers are typically operated. According to Katofsky [7], high temperatures do not lead to higher yield of methanol. The inlet stream is heated by the outlet stream up to the temperature of the reformer to match its heat demand and supply. A typical steam to hydrocarbon ratio of 2:1 for feeds with CO_2 recycling, and 3:1 for feeds without CO_2 recycling are usually maintained. Nonetheless, higher steam to carbon ratio favors a higher H_2:CO ratio and accordingly methanol production. It only implies that more steam would then need to be risen to the reaction temperature.

Methanol synthesis

Via hydrogenation of CO and CO_2 over a suitable catalyst (e.g., copper oxide, zinc oxide, or chromium oxide), methanol is produced.

CO Hydrogenation $CO + 2H_2 \rightleftharpoons CH_3OH, \quad \Delta H = -100.46 \dfrac{kJ}{mol}$

CO_2 Hydrogenation $CO_2 + 3H_2 \rightleftharpoons CH_3OH + H_2O, \quad \Delta H = -61.59 \dfrac{kJ}{mol}$

The stoichiometry of both reactions is satisfied when S in the following relation is 2.03 at least [7]. The outlet stream is conveyed to distillation columns to separate the methanol from waste water.

5.1.4 CO_2 Hydrogenation

Heavy industries such as cement, iron and steel, paper and pulp and refineries have inherent CO_2 emissions within their plant designs, when alternative green means are often considered very expensive due to process high energy demand. This is why these industries find it especially difficult to switch away from fossil fuels. However, carbon capture and utilization (CCU), emerges as a transformative avenue for climate change mitigation and adaptation, offering promising trajectory for these industries to fortify their economic platforms upon widespread commercial applications. Not only their emissions would be reduced, captured CO_2 can be used as a raw material for other processes upon CCU installment. This includes the synthesis of chemicals and materials—in this case methanol (other examples formic acid, polyols for polyurethanes, carbonates), fuels (e.g., methane and kerosene) and direct use in applications based on CO_2 physiochemical properties [8]. Figure 5.4 illustrates the methanol production process utilizing carbon dioxide (CO_2) captured from flue gas as a feedstock. The CO_2 capture unit extracts CO_2, which is then fed into a photocatalytic reactor, where water (H_2O) is used to generate hydrogen (H_2) and oxygen (O_2) through a photocatalytic reaction. The resulting gases pass through a separation unit, where CO_2 is directed toward crude methanol synthesis, while H_2 and CO_2

5.1 Methanol Production

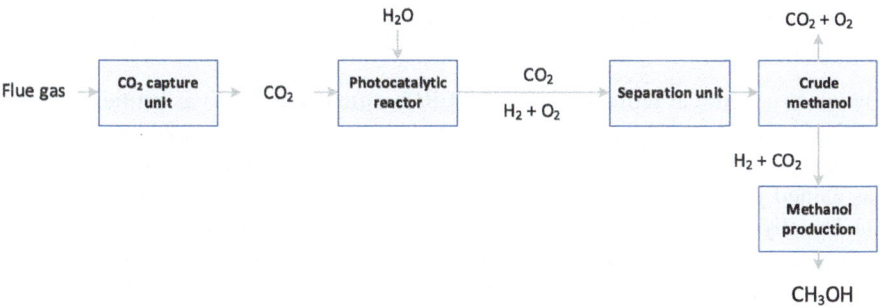

Fig. 5.4 Methanol production from captured CO_2 using photocatalysis

undergo methanol production to form methanol (CH_3OH). This process represents a sustainable approach to CO_2 utilization, contributing to carbon recycling and green methanol production.

Methanol synthesis typically depends on the hydrogenation of carbon monoxide (CO) and carbon dioxide (CO_2). When either of these gases—particularly captured CO_2—is readily available, methanol production becomes more efficient, bypassing several steps of the conventional process.

CO_2 Hydrogenation $\quad CO_2 + 3H_2 \rightleftharpoons CH_3OH + H_2O, \quad \Delta H_{298} = -49.16 \dfrac{\text{kJ}}{\text{mol}}$

Such an approach is typically found in plants that promote environmental sustainability in which energy supply is conducted by renewable means (e.g., solar panels, and wind turbines) as well as its reactant (H_2).

Captured CO is mixed with a H_2 stream, produced by water electrolysis, and preheated before entering the steam raising convertor (SRC) reactor. In the SRC, CO_2 hydrogenation occurs and methanol is produced as well as by product water and CO as a result of reverse WGS reaction.

$$CO_2 + H_2 \rightleftharpoons CO + H_2O, \quad \Delta H_{298} = +41.21 \dfrac{\text{kJ}}{\text{mol}}$$

The produced CO reacts with H_2 for enhanced production of methanol.

CO Hydrogenation $\quad CO + 2H_2 \rightleftharpoons CH_3OH, \quad \Delta H_{298} = -90.77 \dfrac{\text{kJ}}{\text{mol}}$

The reactor outlet stream consisting of methanol, water, and unreacted CO_2, CO, and H_2, thereby, requires separation. Firstly, the stream is condensed and conveyed to a flash separator where unreacted gas is separated from liquid methanol and water due to their dew point differences. As a result, the unreacted gases are pressurized and recycled back to the reactor to enhance the methanol yield. Whereas liquid methanol and water are passed to the distillation column to finally be separated.

5.1.5 Electrochemical CO_2 Reduction

The electrochemical reduction of CO_2 (ERC), often referred to as artificial photosynthesis, involves applying electrical energy to create a potential between two electrodes, enabling the conversion of CO_2 into reduced forms. This technology has gained global interest due to its promising environmental and economic benefits. Usually coupled with renewable energy sources, this technology could generate carbon neutral fuels or industrial chemicals, such as methanol, that are conventionally derived from fossil fuels.

ERC enables the conversion of CO_2 into valuable products using any type of electricity, including renewable sources. In this process, water (H_2O) is oxidized at the anode, producing oxygen (O_2) and releasing electrons (e^-) that travel through an external circuit to the cathode. Simultaneously, protons (H^+) pass through an aqueous electrolyte to reach the cathode. At the cathode, electrons combine with protons or reduce CO_2, resulting in the formation of products such as methanol (CH_3OH), methane (CH_4), and formic acid ($HCOOH$). The distribution of electrons between hydrogen production and CO_2 reduction depends on the activation energy required for each reaction. Below are the possible chemical reactions involved. The following chemical equations are all possible reaction pathways (vs. the standard H_2 electrode (SHE), at 25 °C and pH $= 7$).

$$CO_2 + e^- \rightarrow CO_2^{\cdot -}, \quad E^0 = -1.90\,\text{V}$$
$$CO_2 + 2H^+ + 2e^- \rightarrow HCOOH, \quad E^0 = -0.61\,\text{V}$$
$$CO_2 + 2H^+ + 2e^- \rightarrow CO + H_2O, \quad E^0 = -0.52\,\text{V}$$
$$CO_2 + 4H^+ + 4e^- \rightarrow CH_2O + H_2O, \quad E^0 = -0.51\,\text{V}$$
$$CO_2 + 6H^+ + 6e^- \rightarrow CH_3OH + H_2O, \quad E^0 = -0.38\,\text{V}$$
$$CO_2 + 8H^+ + 8e^- \rightarrow CH_4 + 2H_2O, \quad E^0 = -0.24\,\text{V}$$
$$2H^+ + 2e^- \rightarrow H_2, \quad E^0 = -0.42\,\text{V}$$

To selectively drive the conversion of CO_2 into methanol over other reactions, the use of electrocatalysts is inevitable. During the past decade, several studies have investigated using various solvent phase techniques, including gaseous, aqueous, and non-aqueous, with variety of catalysts, electrode potentials, CO_2 flow rates, pressures, temperature ranges, and reactor cell designs. Yet, more work is required due to the complex reaction pathway of methanol formation. Table 5.1 outlines the major challenges associated with electrochemical CO_2 reduction, which hinder efficiency and scalability. Key issues include high overpotential requirements due to the stability of CO_2, lack of efficient and stable catalysts, and competition with hydrogen evolution, which lowers product selectivity. Additionally, low CO_2 solubility in aqueous solutions and mass transfer limitations further restrict reaction rates and current densities. Proposed solutions involve the use of gas diffusion electrodes, organic solvents, and flow cell designs to enhance CO_2 availability and improve process efficiency.

5.2 Methanol Storage

Table 5.1 Key challenges in electrochemical CO_2 reduction

Challenge	Description
High overpotential requirement	CO_2 is highly stable, requiring a high potential (1.9V vs. SHE) to activate it into the CO_2 anion radical. This leads to significant energy demands
Lack of efficient and stable catalysts	Many possible reaction pathways make it difficult to find catalysts that selectively produce the desired product, avoid side reactions, and last long
Simultaneous H_2 evolution	The potentials for H_2 evolution and CO_2 reduction are similar, causing competition. This reduces product yield and wastes energy by producing H_2 instead of reducing CO_2
Low CO_2 solubility in aqueous solutions	CO_2 solubility in water is low (0.0342 M at room temperature and pressure) and decreases further in high ionic strength solutions, limiting reaction rates.
Mass transfer limitations	Limited CO_2 solubility and diffusion affect current densities. Strategies to mitigate this include gas diffusion electrodes, organic solvents, and flow cell designs [9]

Titanium dioxide (TiO_2) is a widely used photoelectrocatalyst due to its high chemical stability, cost-effectiveness, and low toxicity compared to other metal oxides. However, its large band gap (~ 3.2 eV) and high electron–hole recombination rate limit its efficiency, making it less effective under visible or UV light. To enhance the performance of TiO_2-based photoelectrocatalysts for the photoelectrochemical reduction of CO_2 to methanol, several strategies have been explored:

- Co-catalyst Addition: Incorporating materials like Cu, Ni, Ag, Pd, and graphene to improve charge separation and catalytic efficiency.
- Multi-semiconductor Systems: Combining TiO_2 with other semiconductors and electrocatalysts (e.g., GaP/TiO_2, $TiO_2/NsCuO$, Ti/TiO_2NT-ZrO_2, Cu_2O/TiO_2, and $CuInSe_2/Ni_3Al+TiO_2$) to adjust the band gap and enhance energy conversion.
- Nanostructuring: Developing TiO_2 nanostructures to boost surface area and catalytic activity.

Other metal oxide-based photoelectrodes, such as Cu_2O, In_2O_3, Fe_2O_3, SnO_2, and GaP, have also been investigated. However, TiO_2 remains a focus due to its superior overall performance.

5.2 Methanol Storage

Methanol, a chemical that is both versatile and widely used, necessitates safe and effective storage solutions due to its toxic and flammable properties. Industries and facilities that handle methanol must comprehend the diverse storage options available. Methanol can be stored using physical, chemical, and materials-based methods, each of which has its own advantages and applications. This section discusses the specifics of these three primary storage options. Table 5.2 categorizes methanol storage methods into three primary types: physical storage, chemical storage, and materials-based storage, each with specific sub-types and examples. Physical storage includes liquid storage in above-ground tanks, underground tanks, intermediate bulk containers (IBCs), drums, barrels, and pressurized cylinders, as well as cryogenic storage in double-walled, vacuum-insulated cryogenic tanks. Chemical storage involves methanol reforming systems, which store methanol in chemically bound forms, and conversion to dimethyl ether (DME) as an alternative storage medium. Materials-based storage utilizes adsorption onto porous materials such as activated carbon and metal-organic frameworks (MOFs), as well as absorption in liquid carriers like hydrogels, which allow controlled methanol release through temperature or pressure changes. These diverse storage methods enable safe and efficient handling, transportation, and utilization of methanol in various industrial and energy applications.

5.2.1 Physical Storage

5.2.1.1 Liquid Storage

Liquid storage involves storing methanol in its liquid state at ambient or slightly elevated pressures and temperatures. This method is frequently employed because of its simplicity of implementation and its adaptability to both small and large quantities. There are numerous types of tanks and containers that are employed for liquid storage:

- Above-Ground Storage Tanks: These cylindrical or spherical tanks are intended for industrial use and are capable of accommodating substantial quantities of methanol. In order to mitigate spills and leaks, they are outfitted with secondary containment systems, vents, and pressure relief valves.
- Underground Storage Tanks: Buried tanks are less susceptible to external conditions, which reduces the risk of fire and the environmental impact. Additionally, these tanks are constructed with corrosion-resistant materials and leak detection systems to guarantee their durability and safety.
- Intermediate Bulk Containers (IBCs): IBCs are portable containers that can accommodate moderate quantities of methanol. They are frequently employed in distribution and smaller industrial facilities as a result of their simplicity in transportation and handling.
- Barrels and Drums: These containers are designed for the transportation and storage of methanol in smaller quantities and are constructed from steel or high-density polyethylene (HDPE). They are particularly well-suited for small-scale storage.

5.2 Methanol Storage

Table 5.2 Classification of methanol storage methods

Type	Sub-type	Examples
1. Physical storage	a. Liquid storage	Above-ground storage tanks: large cylindrical or spherical tanks for industrial use
		Underground storage tanks: buried tanks to reduce fire risks and environmental impact
		Intermediate bulk containers (IBCs): portable containers for moderate quantities
		Drums and Barrels: Small-scale, portable storage options
		Pressurized cylinders: for small quantities under pressure
	b. Cryogenic storage	Cryogenic tanks: double-walled, vacuum-insulated containers
2. Chemical storage	a. Chemical bonds in derivatives	Methanol reforming systems: systems that produce methanol from natural gas or biomass, storing it in a chemically bound state until needed
	b. Conversion to other chemicals	DME: methanol can be converted to DME, which can be stored and then reformed back into methanol
3. Materials-based storage	a. Adsorption in porous materials	Activated carbon: methanol is adsorbed onto high-surface area carbon materials.
		Metal-organic frameworks (MOFs): advanced materials with high storage capacity and controlled release properties
	b. Absorption in liquid carriers	Hydrogels: methanol absorbed in hydrophilic gels that release it upon heating or pressure changes

- Pressurized Cylinders: These cylinders are intended for the storage of methanol under pressure and are constructed from steel or aluminum. They are intended for the storage of small quantities. In order to guarantee secure operation, they are furnished with pressure regulators and safety valves.

5.2.1.2 Cryogenic Storage

In an effort to mitigate vapor pressure and enhance safety, cryogenic storage includes the storage of methanol at extremely low temperatures. This method is less prevalent due to the necessity of specialized equipment; however, it provides substantial safety advantages by reducing the likelihood of vapor release and combustion. Cryogenic tanks are vacuum-insulated and double-walled to maintain the low temperatures necessary for cryogenic storage. They are frequently employed in applications that necessitate the storage of methanol for extended periods without vapor loss.

5.2.2 Chemical Storage

5.2.2.1 Chemical Bonds in Derivatives

The controlled release of methanol can be achieved by storing it in the form of chemical derivatives or within chemical compounds. This method is advantageous for combining methanol storage with chemical processes. For example, methanol reforming systems generate methanol by converting natural gas or biomass and store it in a chemically bonded form. Subsequently, the methanol can be liberated and utilized as required, offering a versatile storage solution that combines production and storage procedures.

5.2.2.2 Conversion to Other Chemicals

Methanol can undergo conversion into alternative chemical states for the purpose of storage and subsequently be reconverted when required. This approach can provide advantages in terms of stability and safety. For example, methanol can undergo a conversion process to produce DME, which is a chemically stable compound that offers improved ease of storage and transportation. When necessary, DME can undergo reforming to convert it back into methanol, which makes it a highly efficient method for storing and transporting energy.

5.2.3 Materials-Based Storage

5.2.3.1 Adsorption in Porous Materials

Adsorption storage is the process of trapping methanol within the pores of solid materials. This approach offers a condensed and potentially more secure storage alternative. For example, methanol is adsorbed onto activated carbon materials with a large surface area. This method is efficacious for regulating the discharge of methanol and can be employed in diverse applications. Moreover, metal-organic frameworks (MOFs) are a class of materials. MOFs, or advanced materials, provide significant storage capacities and precise release properties. These materials are specifically engineered to absorb substantial quantities of methanol and subsequently release it as required, rendering them well-suited for specific applications.

5.2.3.2 Absorption in Liquid Carriers

Absorption storage employs liquid carriers capable of absorbing methanol and subsequently releasing it under specific circumstances. This method is beneficial for applications that necessitate controlled release of methanol. For example, hydrogels have the ability to absorb methanol due to their hydrophilic nature. When subjected to heating or changes in pressure, these gels release the absorbed methanol. This offers a regulated delivery mechanism that can be customized to meet specific requirements.

5.3 Methanol Transport

Methanol, a multifaceted and extensively employed chemical compound, is essential in various industrial applications, serving as a feedstock for chemical synthesis, a solvent, and an alternative fuel. The efficient and secure transportation of methanol is crucial to satisfy global industrial demand. Methanol is primarily transported in its liquid form owing to its stability and convenience of handling under ambient conditions. Nonetheless, alternative approaches, including gaseous and novel solid-state transportation, are utilized to meet particular logistical or operational needs. The subsequent sections delineate the principal methods employed for the conveyance of liquid methanol, emphasizing their benefits, challenges, and safety considerations.

5.3.1 Maritime Transport

The maritime transport of methanol is an essential element of the global supply chain, facilitating the efficient and secure transit of this versatile chemical over extensive distances. Specialized methanol tankers, generally possessing a deadweight tonnage (DWT) greater than 30,000 tons, are the principal vessels employed for this purpose. These tankers are expertly engineered to accommodate the distinct characteristics of methanol, guaranteeing operational efficiency and safety. The cargo tanks of these vessels are fabricated from corrosion-resistant materials, including stainless steel or epoxy-coated carbon steel, to avert degradation and maintain the integrity of the storage system over prolonged durations. The durable design of these tanks is crucial for enduring the pressures of maritime transport, encompassing exposure to severe environmental conditions and the mechanical forces experienced during loading and unloading activities.

Besides their robust construction, methanol tankers are outfitted with sophisticated safety mechanisms to reduce the dangers linked to the transportation of a flammable and potentially hazardous material. Inert gas systems are a standard component of these vessels, engineered to displace oxygen in the cargo tanks and inhibit the creation of explosive environments. Vapor recovery units are installed to capture and recycle methanol vapors, thereby reducing emissions and mitigating the risk of fire or explosion. These safety protocols are essential for guaranteeing the secure management of methanol during the entire transportation process, from loading at the point of origin to unloading at the destination.

The maritime transport of methanol is regulated by strict international standards, particularly the International Maritime Dangerous Goods (IMDG) Code. The IMDG Code, established by the International Maritime Organization (IMO), offers a thorough framework for the secure transportation of hazardous materials, such as methanol [10]. The code delineates requirements for vessel design, cargo handling, documentation, and crew training, guaranteeing the application of uniform safety standards worldwide. Adherence to the IMDG Code is obligatory for all vessels involved in the international transport of methanol, thereby improving the safety and dependability of maritime methanol transportation [11].

For shorter inland routes, smaller barges provide a more economical option compared to ocean-going tankers. These barges, which traverse rivers and canals, are equipped with specialized tanks that include numerous safety features akin to those on larger tankers, albeit in a reduced scale. Although their capacity is markedly inferior to that of ocean-going vessels, barges offer a versatile and effective means of transporting methanol to areas inaccessible to larger ships. Barges are especially beneficial for transporting methanol to industrial sites situated along inland waterways, minimizing the necessity for overland transport and its related expenses.

Notwithstanding the benefits of maritime and inland waterway transport, the sector encounters persistent challenges concerning safety, environmental sustainability, and regulatory adherence. The advancement of sophisticated safety technologies, including real-time monitoring systems and improved spill containment measures, is crucial for tackling these challenges. The implementation of sustainable practices, including low-emission fuels and energy-efficient vessel designs, is gaining significance as the maritime industry aims to minimize its environmental impact.

In summary, the maritime transport of methanol depends on specialized tankers and barges engineered to guarantee the secure and efficient conveyance of this vital chemical. The utilization of corrosion-resistant materials, sophisticated safety systems, and compliance with international regulations, such as the IMDG Code, is essential for preserving the integrity of the supply chain. Although smaller barges offer an economical option for inland routes, continuous improvements in safety and sustainability will be crucial to satisfy the increasing demand for methanol and tackle the challenges of global transportation.

5.3.2 Land Transport

The terrestrial transportation of methanol is a vital component of its global supply chain, guaranteeing the efficient and safe delivery of this essential chemical to industries and consumers. Two principal modes of terrestrial transportation utilized are

5.3 Methanol Transport

tank trucks and rail tank cars. Each method presents unique benefits based on the quantity of methanol transported, the distance covered, and the particular logistical needs of the area.

Tank trucks serve as a highly adaptable method for transporting methanol, especially over shorter distances or in areas devoid of pipeline infrastructure. The vehicles vary in size from small delivery trucks, which can transport several thousand liters, to large articulated lorries with capacities surpassing 30,000 l. The versatility of tank trucks renders them optimal for transporting methanol to designated locations, including industrial facilities, agricultural sites, and remote areas where alternative transportation methods are unfeasible. Their agility facilitates accurate delivery, guaranteeing that methanol arrives at end-users promptly and efficiently.

Tank trucks utilized for methanol transportation are fitted with specialized tanks engineered to securely contain the substance. These tanks are generally fabricated from corrosion-resistant materials, such as stainless steel or aluminum, and are frequently insulated to preserve the stability of methanol during transportation. Moreover, tank trucks are equipped with safety features such as pressure relief valves, vapor recovery systems, and spill containment mechanisms to reduce the risks linked to the transportation of flammable and potentially hazardous materials. Operators of these vehicles must complete specialized training to manage methanol safely, encompassing emergency response protocols and adherence to hazardous materials regulations.

Rail tank cars are the preferred method for transporting substantial volumes of methanol over long distances. These vehicles, capable of transporting up to 30,000 gallons (113,562 l) of methanol, utilize extensive rail networks to offer an economical and dependable distribution method. Rail transport is especially beneficial for transporting methanol from production sites to significant industrial centers or export terminals, where substantial quantities are necessary. Utilizing rail tank cars diminishes the required trips in comparison with tank trucks, consequently reducing transportation expenses and mitigating environmental effects related to fuel consumption and emissions.

Rail tank cars are engineered with stringent safety mechanisms to guarantee the safe transportation of methanol. The tanks are fabricated from high-strength materials, including carbon steel with protective coatings, to avert corrosion and endure the demands of rail transport. They are additionally outfitted with thermal insulation to preserve methanol in its liquid form and avert vaporization. Safety mechanisms, including pressure relief devices and sophisticated braking systems, are incorporated into the design to mitigate potential hazards during transit. Similar to tank truck operators, railway personnel engaged in methanol transportation must comply with stringent safety protocols and undergo specialized training for hazardous materials handling.

The conveyance of methanol via tank trucks and rail tank cars is governed by stringent safety regulations to safeguard human health, property, and the environment. The U.S. Department of Transportation (DOT) enforces the Hazardous Materials Regulations (HMR) in the United States, which delineate extensive requirements for the design, construction, labeling, and operation of vehicles transporting methanol

[12]. In Europe, the European Agreement concerning the International Carriage of Dangerous Goods by Road (ADR) and the Regulation concerning the International Carriage of Dangerous Goods by Rail (RID) set standardized safety protocols for the terrestrial transport of methanol and other hazardous materials [10]. These regulations require the utilization of certified tanks, appropriate labeling and documentation, and comprehensive training programs for drivers and operators.

Notwithstanding the stringent safety protocols implemented, the terrestrial transport of methanol presents challenges. The flammability and toxicity of methanol require ongoing improvements in safety technologies, including superior tank materials, advanced spill containment systems, and real-time monitoring devices. The advancement of sustainable transportation methods, encompassing alternative fuels and energy-efficient vehicles, is essential for mitigating the environmental impact of methanol transport.

In summary, the terrestrial transportation of methanol using tank trucks and rail tank cars is a meticulously regulated and efficient procedure that guarantees the secure delivery of this essential chemical to diverse industries. Tank trucks provide flexibility and accessibility for localized distribution, whereas rail tank cars offer a cost-effective solution for extensive, long-distance transportation. Compliance with rigorous safety regulations and continuous innovation in transportation technologies are crucial for preserving the integrity of the methanol supply chain and tackling the challenges of terrestrial transport.

5.3.3 Pipelines

Pipelines are among the most efficient and environmentally sustainable means of transporting methanol, especially for large-scale and continuous distribution across long distances. These pipelines, fabricated predominantly from premium steel, are designed to endure the pressures necessary for the secure and effective transportation of methanol. Pipelines reduce the necessity for intermediate storage and handling, thereby diminishing the risk of spills and emissions linked to alternative transportation methods. Pipelines exhibit a reduced carbon footprint relative to road or rail transport, as they utilize less energy-intensive methods for transporting liquids over extensive distances. The establishment of methanol pipelines is limited by substantial initial capital expenditures and possible environmental issues, including land disturbance and habitat disruption during construction. Methanol pipelines are primarily employed in areas with robust industrial infrastructure or significant consumption centers, where the economic advantages of a consistent and dependable supply chain warrant the initial expenditures.

Although liquid methanol is the predominant form utilized for transportation, gaseous methanol is employed in particular industrial applications or for short-distance conveyance. Gaseous methanol is generally transported in compressed gas cylinders or tube trailers, specifically engineered to securely contain the pressurized gas. This technique is especially beneficial for sectors that necessitate methanol in

5.3 Methanol Transport

gaseous state for chemical synthesis or as a feedstock for specific processes. The transport of gaseous methanol requires rigorous safety measures due to the inherent dangers of pressurized gases, such as the risk of leaks, explosions, or asphyxiation. Specialized handling protocols, routine equipment inspections, and sophisticated monitoring systems are imperative for the secure transportation of gaseous methanol. This method is less common than liquid transportation due to greater costs and logistical complexities, despite its specialized applications.

Solid-state methanol, including methanol-impregnated activated carbon and methanol-based chemical hydrides, represents a burgeoning field of research that offers potential for safer and more convenient transportation and storage. This method entails the adsorption of methanol onto porous materials or its chemical binding into stable solid compounds, which can subsequently release methanol as required. Solid-state methanol presents numerous advantages compared to liquid methanol, such as diminished spill risks, decreased volatility, and enhanced handling and storage convenience. These attributes render it especially appealing for use in remote or off-grid areas, where conventional transportation methods may be unfeasible. Moreover, solid-state methanol could improve safety in urban or densely populated regions by reducing the risks linked to liquid methanol, including vapor emissions and flammability. Nonetheless, the technology for solid-state methanol transportation remains in its nascent phase of development. Extensive research and investment are essential to enhance the adsorption and desorption processes, augment the stability and capacity of the utilized materials, and establish the requisite infrastructure for large-scale deployment. Issues including material degradation across numerous cycles, energy demands for methanol liberation, and economic feasibility must be resolved prior to the widespread adoption of solid-state methanol.

The selection of transportation method for methanol is determined by a confluence of environmental, economic, and logistical considerations. Pipelines, although efficient and sustainable, are constrained by substantial initial costs and environmental repercussions during construction. Transporting gaseous methanol, while appropriate for certain applications, is less scalable due to safety issues and elevated operational expenses. Solid-state methanol, although promising, necessitates additional advancement to surmount technical and economic obstacles. Each method possesses distinct advantages and challenges, and the optimal selection is contingent upon the specific requirements of the supply chain, including distance, volume, and end-use applications.

Continuous research and innovation are essential for the advancement of methanol transportation technologies and for overcoming current limitations. Innovations in materials science and construction methodologies may diminish expenses and environmental repercussions for pipelines. The advancement of safer and more efficient compression and storage systems for gaseous methanol could enhance its utility. Advancements in material design and process optimization could expedite the commercialization of solid-state methanol. The incorporation of renewable energy sources and sustainable practices into methanol transportation systems is crucial for minimizing the overall environmental impact of this essential industrial process.

Table 5.3 Overview of transportation modes for methanol: advantages and disadvantages

	Mode of transportation	Advantages	Disadvantages
Liquid	Maritime (tankers)	High capacity, suitable for long distances, relatively low cost per unit	High initial investment, potential for marine pollution in case of spills, limited accessibility to inland locations
	Maritime (barges)	Cost-effective for inland waterways, flexible	Limited capacity, restricted to navigable waterways, slower than rail or pipeline
	Land (rail tank cars)	Efficient for bulk transport over land, well-established infrastructure, relatively safe	Limited flexibility, potential for accidents and spills, requires access to rail network
	Land (tank trucks)	Versatile, suitable for localized deliveries, flexible routing	Lower capacity, higher cost per unit compared to rail or pipeline
	Land (pipelines)	Cost-effective for long distances, minimal environmental impact during operation, high capacity	High initial investment, potential for leaks and land disruption, limited existing infrastructure
Gaseous	Cylinders/tube trailers	Suitable for smaller volumes and specific applications, easier handling than liquid methanol	Lower capacity, higher cost per unit, requires specialized handling equipment and safety measures
Solid	Methanol-impregnated activated carbon)	Potentially safer and more convenient for storage and handling, reduced risk of spills	Requires additional infrastructure and processes for methanol extraction, technology still under development

In conclusion, the transportation of methanol involves various methods, each presenting distinct advantages and challenges. Table 5.3 provides an overview of the different transportation modes for methanol, highlighting their advantages and disadvantages. Additionally, the selection of the most suitable method for transporting methanol depends on several factors, such as the quantity of methanol to be transported, the distance to be covered, the existing infrastructure, the cost, and the considerations of safety and environmental impact. Tankers are the preferred mode of transportation for long-distance and high-capacity maritime shipping, whereas tank trucks and rail tank cars are better suited for distributing goods on land. Pipelines provide a highly effective method for maintaining a continuous supply but necessitate a substantial initial investment. Pipelines present an efficient and sustainable method for large-scale transportation, whereas gaseous and solid-state methanol serve as specialized alternatives for particular applications. The ongoing advancement of these technologies, bolstered by thorough research and investment, will be crucial for satisfying the increasing global demand for methanol while maintaining safety, efficiency, and environmental sustainability.

References

1. Dalena F, Senatore A, Marino A, Gordano A, Basile M, Basile A (2018) Methanol production and applications: an overview. Methanol Sci Eng 3–28. https://doi.org/10.1016/B978-0-444-63903-5.00001-7
2. Rozovskii AY, Lin GI (1999) Catalytic synthesis of methanol. Kinet Catal 40:773–794
3. Rabenhorst DW (1975) Methanol from coal-a step toward energy self-sufficiency. Energy Sour 2:251–262. https://doi.org/10.1080/00908317508945952
4. Argonul A, Er OO, Kayahan U, Unlu A, Ziypak M (2021) Syngas cleaning for coal to methanol demo plant—H_2S and COS removal. Chem Eng Commun 208:950–964. https://doi.org/10.1080/00986445.2020.1722653
5. Hamelinck CN, Faaij APC (2019) Production of methanol from biomass. In: Alcohol fuels, pp 7–50. https://doi.org/10.1201/9781420020700-3
6. Pierik JTG, Curvers APWM (1995) Logistics and pre-treatment of biomass fuels for gasification and combustion
7. Katofsky RE (1993) The production of fluid fuels from biomass
8. Pérez-Fortes M, Schöneberger JC, Boulamanti A, Tzimas E (2016) Methanol synthesis using captured CO_2 as raw material: techno-economic and environmental assessment. Appl Energy 161:718–732. https://doi.org/10.1016/j.apenergy.2015.07.067
9. Wiranarongkorn K, Eamsiri K, Chen YS, Arpornwichanop A (2023) A comprehensive review of electrochemical reduction of CO_2 to methanol: technical and design aspects. J CO_2 Util 71. https://doi.org/10.1016/j.jcou.2023.102477
10. IMO (2020) International maritime dangerous goods (IMDG) code. London
11. Alliance Consulting International (2008) Methanol safe handling manual, vol 5
12. DOT (2023) Code of federal regulations, title 49: transportation of hazardous materials. Washington, D.C

Open Access This chapter is licensed under the terms of the Creative Commons Attribution 4.0 International License (http://creativecommons.org/licenses/by/4.0/), which permits use, sharing, adaptation, distribution and reproduction in any medium or format, as long as you give appropriate credit to the original author(s) and the source, provide a link to the Creative Commons license and indicate if changes were made.

The images or other third party material in this chapter are included in the chapter's Creative Commons license, unless indicated otherwise in a credit line to the material. If material is not included in the chapter's Creative Commons license and your intended use is not permitted by statutory regulation or exceeds the permitted use, you will need to obtain permission directly from the copyright holder.

Chapter 6
Dimethyl Ether

6.1 DME Production

Dimethyl Ether (DME) refers to a chemical compound (CH_3OCH_3) that is derived from methanol (CH_3OH), and has recently gained attention for its use as a fuel in efficient engines, as well as for its ability to lower emissions of NO_x and SO_x. DME is the simplest ether with no C-C bond. At atmospheric conditions, DME is a gas, but due to its vapor pressure of 5.1 bar and temperature of 20 °C, it can be easily liquefied, stored, and transported in pressurized tanks [1]. In principle, DME can be produced in two distinct ways: (1) the indirect route uses the produced methanol to promote its dehydration, (2) the direct route, arguably deemed to be commercially unattractive, in which DME is produced in a single stage using bifunctional catalysts.

The variety of DME synthesis comes down to methanol/syngas origins in which the two-step processes involving the conversion of syngas and the subsequent dehydration of methanol could then be carried out. Figure 6.1 illustrates the DME production process, highlighting both non-renewable (natural gas, oil, and coal) and renewable sources (water, sunlight, heat, and biomass). Non-renewable pathways involve steam methane reforming, methane pyrolysis, partial oxidation, plasma reforming, and gasification, while renewable pathways use electrolysis, photobiological, photoelectrochemical, thermochemical, and biomass gasification techniques. The resulting syngas and hydrogen (H_2) with carbon dioxide (CO_2) are utilized in methanol synthesis, which then undergoes DME synthesis to produce DME, a cleaner alternative to conventional fuels with applications in transportation, power generation, and industrial processes.

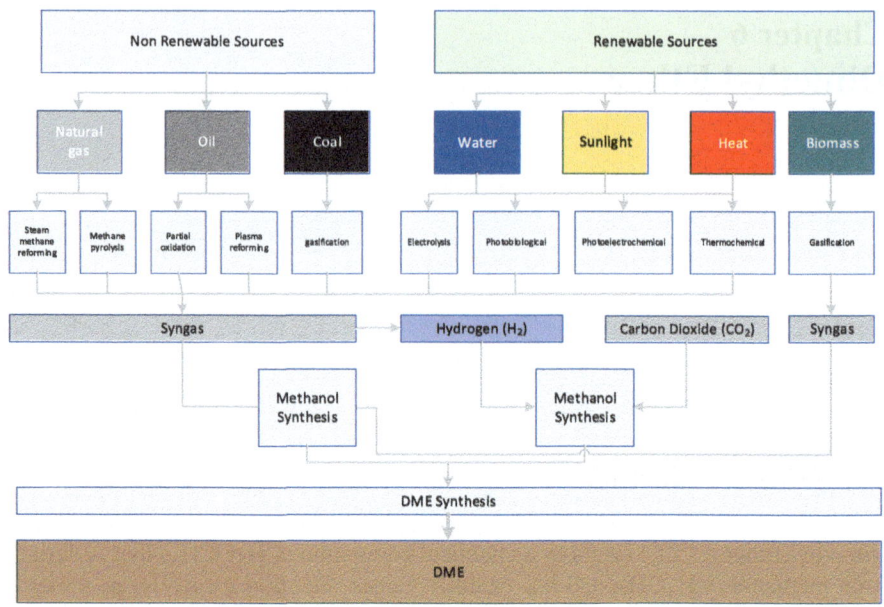

Fig. 6.1 DME production pathways from non-renewable and renewable sources

6.1.1 Methanol Dehydration

DME production via methanol dehydration follows the indirect route in which natural gas is first used to synthesize and purify methanol, which is then converted to DME and water. This route is very well-developed and commercially established and possibly the most efficient one [2]. Figure 6.2 illustrates the DME production process using natural gas as a feedstock. The process begins with desulfurization to remove impurities, followed by steam reforming, where natural gas reacts with air to produce synthesis gas (syngas) containing hydrogen (H_2) and carbon monoxide (CO). The syngas then undergoes DME and methanol synthesis, converting it into DME and methanol intermediates. A final purification unit removes excess water (H_2O), yielding pure DME, which is a clean-burning fuel alternative used in transportation, power generation, and aerosol propellants.

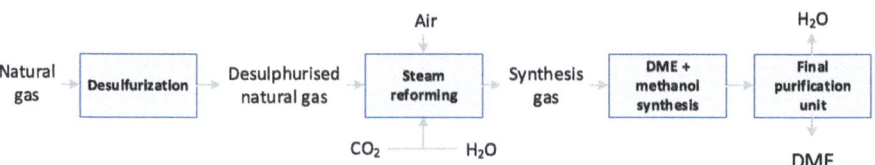

Fig. 6.2 DME production from natural gas

6.1 DME Production

The methanol dehydration reaction typically performed adiabatically in the vapor phase at 1–20 bar, with an inlet temperature range of 180–275 °C and a maximum outlet temperature of 400 °C in the presence of gamma-alumina catalyst.

$$2CH_3OH \rightarrow CH_3OCH_3 + H_2O, \quad \Delta H^0_{298} = -23.5 \frac{kJ}{mol}$$

Reported in [3], the fresh pressurized methanol combined with a recycle stream composed of unreacted methanol and water traces are fed into the preheating section. The steam is first vaporized at 154 °C; then the vapor is preheated to the reactor inlet temperature of 250 °C using the high-temperature adiabatic reactor effluent [4]. Owing to the exothermicity of the reaction, the reactor effluent leaves at 371 °C. It is important to note a critical limitation on the reactor temperature is that the deactivation temperature of the gamma-alumina catalyst is 400 °C. Next, the reactor effluent is cooled to 90 °C and conveyed to the first distillation column which separates DME from methanol and water, and the second methanol from water at 99% purities.

6.1.2 Direct DME Synthesis from Syngas

Direct DME synthesis combines both processes of methanol production and dehydration in one reactor at large-scale industrial plants. Such approach requires the embodiment of a dual catalyst system that acts as methanol synthesis catalyst and a methanol dehydration catalyst, immersed in a single unit.

$$3CO + 3H_2 \rightleftharpoons CH_3OCH_3 + CO_2, \quad \Delta H^0 = -245.7 \frac{kJ}{mol}$$

$$2CO_2 + 6H_2 \rightleftharpoons CH_3OCH_3 + 3H_2O, \quad \Delta H^0 = -204.8 \frac{kJ}{mol}$$

Within these overall reactions, it consists of three other reactions. All of which are prerequisites of the first reaction (CO hydrogenation): methanol synthesis, dehydration of methanol, and a water gas shift (WGS) reaction, shown below in the same order. In the absence of WGS, direct DME synthesis can be carried out by the second reaction (CO_2 hydrogenation).

$$CO + 2H_2 \rightleftharpoons CH_3OH, \quad \Delta H^0 = -90.7 \frac{kJ}{mol}$$

$$2CH_3OH + 2H_2 \rightleftharpoons CH_3OCH_3 + H_2O, \quad \Delta H^0 = -23.7 \frac{kJ}{mol}$$

$$CO + H_2 \rightleftharpoons CO_2 + H_2O, \quad \Delta H^0 = -40.9 \frac{kJ}{mol}$$

The composition of the inlet syngas determines the dominant reaction pathway. For traditional feedstocks, such as natural gas, CO hydrogenation is the primary reaction. Conversely, for CO_2-rich feedstocks—derived from carbon capture or biomass gasification—CO_2 hydrogenation becomes the prevailing process. Additionally, the pathways can be linked via reverse WGS (RWGS) reaction. RWGS is preferred since it is significantly less exothermic than WGS, being less critical for the heat management. Nonetheless, the extremely high heat released by the abovementioned exothermic reactions poses a critical challenge related to the heat transfer in the reactor design.

Due to the drastic difference in feasibility and technology readiness when compared to indirect synthesis, direct DME synthesis is mostly found in lab experimentation and simulation, and pilot scale plants.

Direct DME synthesis was investigated [5], in a stainless steel fixed bed reactor with an inner diameter of 12 mm and a length of 460 mm, filled with a physical mixture of 2 g admixed catalyst and 20 g SiC. In four independent heating zones, the reactor was heated to ensure an axial temperature difference within the catalyst bed of typically less than 2 °C. Pure feed gases included in CO, Ar, N_2, and H_2, and a 50:50 mixture of CO_2 and N_2, were controlled using mass flow controls, to produce DME.

The study highlights the exceptional performance of ferrierite-type zeolites (FER) in direct DME synthesis from variable CO_2/CO_x feeds, outperforming g-Al_2O_3 due to superior catalytic activity and water resistance. Pairing Cu-Zn-Zr (CZZ) with FER resulted in significantly higher DME productivity and CO_x conversion compared to Cu-Zn-Al (CZA) systems, under both CO-rich (1017 g DME (kgCu h)$^{-1}$) and CO_2-rich (689 g DME (kgCu h)$^{-1}$) conditions at 523 K.

The CZZ/FER system maintained high selectivity (~ 90%) and operational flexibility across different CO_2/CO_x ratios and temperatures, with only a 27% productivity decline over 545 h. These findings position CZZ/FER as a promising catalyst for sustainable power-to-fuel technologies, utilizing renewable hydrogen and captured CO_2 to drive efficient, flexible DME production [5].

6.1.3 Biomass Gasification

Topping the chemical and physical properties of DME: having higher cetane number, reduced NO_x and SO_x-enabling it to be an easy substitute to liquefied petroleum gases LPGs (e.g., diesel), with carbon neutral fuel origins would be a game changer in the DME applications. The CO_2 generated from DME derived from biomass can be offset naturally by photosynthesis. While the idea is extremely attractive, not much work has been performed to study DME production using biomass as a feedstock albeit the commercial availability of fossil fuel gasification for DME production. This is due to the low H_2:CO ratio and high CO_2 concentration in syngas pose a major challenge in biomass-based DME synthesis [6].

6.1 DME Production

CO Hydrogenation $CO + 2H_2 \rightleftharpoons CH_3OH$, $\Delta H = -90.77 \frac{kJ}{mol}$

CO$_2$ Hydrogenation $CO_2 + 3H_2 \rightleftharpoons CH_3OH + H_2O$, $\Delta H = -49.16 \frac{kJ}{mol}$

WGS $CO + H_2O \rightarrow CO_2 + H_2$, $\Delta H = -42 \frac{kJ}{mol}$

DME synthesis $2CH_3OH + 2H_2 \rightleftharpoons CH_3OCH_3 + H_2O$, $\Delta H = -23.7 \frac{kJ}{mol}$

Making the overall reaction

$$2CO + 2H_2 \rightleftharpoons CH_3OCH_3 + CO_2, \quad \Delta H = -205.63 \frac{kJ}{mol}.$$

The process could conceptually mirror the schematic pathway of fossil fuel gasification used for bio-based methanol, with adjustments to facilitate methanol conversion into DME. Alternatively, a more efficient, single-step approach could also be implemented directly. Figure 6.3 illustrates the DME production process from biomass as a renewable feedstock. The process begins with gasification, where biomass is converted into raw syngas (a mixture of CO, H$_2$, and other gases). The water gas shift reaction adjusts the syngas composition by adding hydrogen. The syngas then undergoes purification to remove impurities such as water (H$_2$O), hydrogen sulfide (H$_2$S), and carbon dioxide (CO$_2$). The purified syngas is then used in DME synthesis, followed by a final purification unit to remove residual methanol (MeOH) and water, yielding pure DME. This process supports sustainable fuel production with applications in transportation, power generation, and industrial energy systems.

Simulated in [7], a single-step DME production based on conventional biomass (i.e., rice straw was selected as the biomass feedstock with the tabulated basic properties shown below) consists of 5 main units: gasification, WGS, purification, a single-step DME synthesis, and DME separation. Fluidized-bed gasifiers for biomass gasification typically operate at 750–1100 °C with an oxidizing agent/biomass ratio of 0.3–0.5. This study conducted sensitivity analysis to determine the influence of

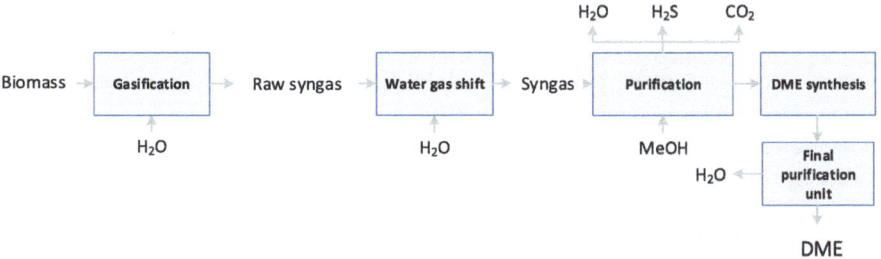

Fig. 6.3 Biomass-based DME production process

operating conditions on syngas composition, aiming for an H_2/CO ratio of 1 and 3 mol% CO_2 [8].

At 1000 kg/h, the biomass was gasified at 950 °C and 0.1 MPa, with steam (150 kg/h) and CO_2 (345 kg/h) introduced. The syngas exiting at 950 °C and 0.1 MPa was cooled to 250 °C, compressed to 5 MPa, and purified using a ZnO-based process, reducing H_2S to below 1 ppm.

The purified syngas was then fed into the DME synthesis reactor at 250 °C and 5 MPa, where a Cu-based bifunctional powder catalyst facilitated the reaction. The resulting gas mixture was cooled to 40 °C, condensing DME, CO_2, and water, which were separated via flash separation at 3 MPa. Non-condensable gases were removed, and the condensate was distilled in a 10-tray column at 3 MPa, yielding a bottom product with 97.3 mol% DME and a top product with 83.5 mol% CO_2. A portion of the CO_2 was recycled to the gasifier as an oxidizing agent.

6.1.4 DME Production from Captured CO_2 and Green Hydrogen

The technologies employed for methanol production through captured CO_2 hydrogenation (i.e., CCU) can, in principle, be seamlessly adapted for the synthesis of DME. Adopting the indirect route, the synthesized methanol could be conveyed into a dehydration unit to convert the intermediate product into DME.

CO_2 Hydrogenation : $CO_2 + 3H_2 \rightleftharpoons CH_3OH + H_2O$, $\Delta H_{298} = -49.16 \frac{kJ}{mol}$
Considering RWGS : $CO_2 + H_2 \rightleftharpoons CO + H_2O$, $\Delta H_{298} = +41.21 \frac{kJ}{mol}$
The produced CO reacts with H_2 for enhanced production of methanol.
CO Hydrogenation : $CO + 2H_2 \rightleftharpoons CH_3OH$, $\Delta H_{298} = -90.77 \frac{kJ}{mol}$
Methanol dehydration : $2CH_3OH \rightarrow CH_3OCH_3 + H_2O$, $\Delta H_{298}^0 = -23.5 \frac{kJ}{mol}$
Taking the global reaction leading to the formation DME:

$$2CO_2 + 6H_2 = CH_3OCH_3 + 3H_2O, \quad \Delta H_{298}^0 = -122.22 \frac{kJ}{mol}.$$

To maximize the industrial effort of combating climate change, the hydrogen used for the hydrogenation reactions could be obtained from water electrolysis supplied by renewable energy. By pursuing this strategy and intensifying research on the outlined steps, it will be possible to establish an efficient CO_2-based production system for both chemicals and energy. This approach balances energy demands—reducing reliance on fossil fuels—while enhancing socio-environmental sustainability by curbing CO_2 emissions into the atmosphere. Despite significant progress in recent years, developing a highly efficient catalyst for CO_2 hydrogenation remains a compelling challenge. The direct synthesis of DME through one-step CO_2 hydrogenation involves two key reactions: methanol formation and its subsequent dehydration. Consequently, the catalyst must possess a redox function capable of hydrogenating CO_2 to alcohol, alongside an acid function to facilitate the alcohol-to-ether conversion. On the other hand, DME indirect synthesis is well-established offering a variety of large-scale production routes.

6.2 DME Storage

DME is a versatile fuel and chemical precursor recognized for its advantageous characteristics, including a high cetane number, efficient combustion, and the capacity to be stored as a liquid at moderate pressure. DME storage can be categorized into three main types: physical storage, chemical storage, and material-based storage. Figure 6.4 categorizes DME storage into three main types: physical storage, chemical storage, and material-based storage. Physical storage includes pressurized storage in vessels or tanks and cryogenic storage at extremely low temperatures to maintain DME in liquid form. Chemical storage involves hydrates or clathrates, DME-methanol systems, and DME derivatives that undergo reversible chemical reactions for controlled release. Material-based storage utilizes adsorption and absorption techniques, including activated carbon, zeolites, metal-organic frameworks (MOFs), ionic liquids, and polymers, which enhance safe and efficient DME containment. These methods enable optimized storage and transport of DME for energy and industrial applications.

6.2.1 Physical Storage

DME, a multifaceted chemical compound utilized as a clean-burning fuel, aerosol propellant, and chemical feedstock, necessitates particular storage conditions to preserve its liquid state for effective handling and transportation. The two principal techniques for storing DME in its liquid state are pressurized storage and cryogenic storage. Each method possesses unique benefits and challenges, rendering them appropriate for various applications based on considerations such as storage duration, volume requirements, and infrastructure availability.

6.2.1.1 Pressurized Storage

Pressurized storage is the predominant technique for preserving DME in its liquid form. This method utilizes the physical characteristics of DME, which converts to a liquid at moderate pressures between 5 and 10 bar at ambient temperatures. DME can be stored and transported utilizing infrastructure intended for liquefied petroleum gas (LPG), necessitating only minimal modifications. The compatibility with current LPG infrastructure markedly diminishes the expenses and intricacies linked to the implementation of DME as an alternative fuel or industrial chemical [9].

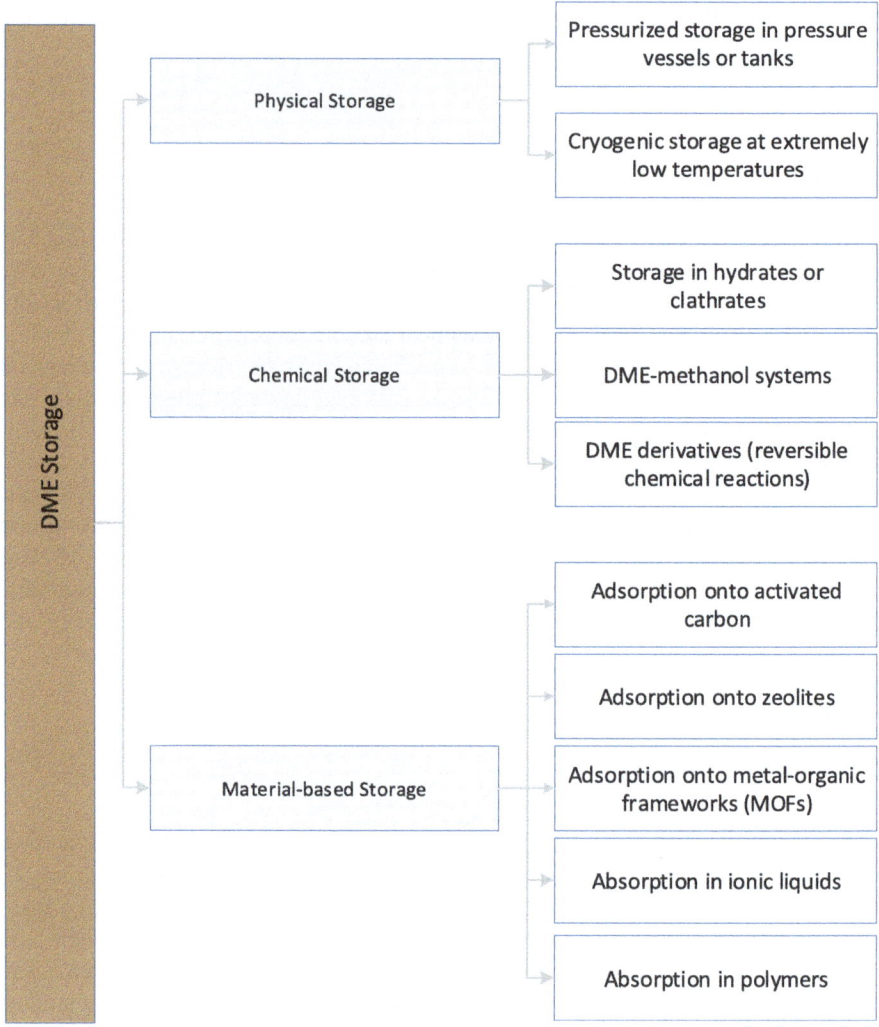

Fig. 6.4 Classification of DME storage methods

Pressure vessels or tanks utilized for DME storage are generally fabricated from high-strength materials, such as carbon steel or stainless steel, to endure the necessary pressures and avert leaks. These vessels are outfitted with safety mechanisms, such as pressure relief valves, temperature monitoring systems, and vapor recovery units, to guarantee secure and efficient storage. Pressurized storage is especially beneficial for applications necessitating regular access to DME, such as fuel distribution systems or industrial operations, owing to its simplicity and cost efficiency.

6.2.1.2 Cryogenic Storage

Cryogenic storage, while less prevalent due to elevated energy demands and intricate infrastructure, presents a feasible option for preserving DME in its liquid form. This technique entails cooling DME to exceedingly low temperatures, generally below − 25 °C, where it remains in a liquid state at atmospheric pressure. Cryogenic storage is especially appropriate for applications necessitating extensive, long-term preservation, such as strategic fuel reserves or bulk transport over considerable distances. Reduced temperatures guarantee the stability of DME for prolonged durations, thereby diminishing the likelihood of vaporization or degradation [10].

The implementation of cryogenic storage requires specialized equipment, such as insulated tanks, refrigeration systems, and temperature control mechanisms, resulting in elevated capital and operational expenses. The energy-intensive requirement for sustaining cryogenic temperatures renders this method economically unfeasible for smaller-scale or short-term storage applications. Notwithstanding these challenges, cryogenic storage continues to be a vital alternative for industries necessitating high-capacity storage solutions with reduced risk of product loss.

The selection between pressurized and cryogenic storage is contingent upon the particular needs of the application. Pressurized storage is typically favored due to its simplicity, reduced costs, and compatibility with existing infrastructure, rendering it suitable for extensive application in fuel distribution and industrial processes. Cryogenic storage, conversely, is designated for specialized applications requiring long-term stability and substantial storage capacities, such as in the aerospace sector or for strategic energy reserves.

Both pressurized and cryogenic storage techniques necessitate rigorous safety protocols to alleviate the hazards linked to DME, which is combustible and can generate explosive mixtures with air. Safety protocols encompass the implementation of resilient containment systems, routine inspections, and sophisticated monitoring technologies to identify leaks or pressure fluctuations. Environmental factors are crucial, as the energy demands of cryogenic storage impact its carbon footprint. Enhancing the energy efficiency of cryogenic systems and advancing renewable energy-powered refrigeration technologies are crucial for mitigating the environmental impact of DME storage.

Current research aims to optimize DME storage technologies to improve safety, efficiency, and sustainability. Advancements in materials science, including the creation of sophisticated composites for pressure vessels, may enhance the durability and efficacy of pressurized storage systems. Advancements in cryogenic technology, such as magnetic refrigeration and phase-change materials, may diminish the energy demands and expenses linked to cryogenic storage. The incorporation of intelligent monitoring systems and automation could further improve the safety and reliability of DME storage infrastructure.

In summary, the storage of DME in liquid form is accomplished via pressurized and cryogenic techniques, each presenting distinct benefits and difficulties. Pressurized storage is the most common method due to its cost efficiency and compatibility with current infrastructure, whereas cryogenic storage is utilized for specialized applications that demand long-term stability and substantial capacities. The ongoing advancement of storage technologies, underpinned by stringent safety measures and environmental factors, will be crucial in addressing the increasing demand for DME as a sustainable fuel and industrial chemical.

6.2.2 Chemical Storage

The chemical storage of DME constitutes a novel method for managing this multifaceted compound by maintaining it in a chemically bonded form, facilitating regulated release as needed. This approach utilizes the principles of chemical bonding and reversible reactions to improve the safety, stability, and efficiency of DME storage. Two principal strategies utilized in chemical storage are the formation of hydrates or clathrates and the implementation of reversible chemical reactions. Each method presents distinct benefits and obstacles, rendering them appropriate for particular applications based on considerations such as storage duration, release mechanisms, and infrastructure needs.

6.2.2.1 Hydrates and Clathrates

A promising approach for the chemical storage of DME entails the formation of hydrates or clathrates, which are crystalline solid compounds wherein DME molecules are encapsulated within a lattice of water molecules. These structures are reinforced by hydrogen bonding and van der Waals forces, enabling DME to be stored at considerably reduced pressures and temperatures relative to its pure liquid or gaseous forms. DME hydrates can be stabilized at temperatures slightly above 0 °C and moderate pressures, thereby minimizing energy demands and safety hazards linked to high-pressure or cryogenic storage [11].

The utilization of hydrates and clathrates presents multiple benefits, such as improved safety resulting from the diminished volatility of DME in this state and the possibility of increased storage densities. The release of DME from these structures can be regulated by modifying temperature and pressure conditions, rendering this method appropriate for applications necessitating precise dosing or gradual release. The formation and dissociation of hydrates and clathrates necessitate meticulous optimization to guarantee efficient storage and retrieval of DME. To render this method commercially viable, challenges, including sluggish formation kinetics, limited storage capacity, and the necessity for meticulous control of environmental conditions, must be resolved.

6.2 DME Storage

6.2.2.2 Reversible Chemical Reactions

An alternative method for chemical storage utilizes reversible chemical reactions, wherein DME is chemically incorporated into a larger molecule or complex. DME can be stored within a DME-methanol system or as a derivative compound, such as methylal (dimethoxymethane). These compounds exhibit stability under ambient conditions and can liberate DME via controlled chemical reactions, including hydrolysis or thermal decomposition, when subjected to particular catalysts or temperature conditions [12].

This method provides multiple advantages, such as enhanced stability and safety during storage, as the DME is chemically immobilized and less susceptible to leakage or inadvertent release. The utilization of reversible reactions facilitates the efficient and regulated release of DME, rendering it appropriate for applications necessitating precise delivery, such as fuel cells or chemical synthesis processes. The development of effective catalysts and reaction conditions for DME release continues to pose a significant challenge. The energy demands for bond cleavage and the potential for byproduct generation must be meticulously regulated to guarantee the feasibility and sustainability of this method.

The selection between hydrate/clathrate storage and reversible chemical reactions is contingent upon the particular demands of the application. Hydrate and clathrate storage are especially beneficial for applications necessitating moderate storage durations and regulated release, such as in portable fuel systems or emergency energy reserves. Conversely, reversible chemical reactions are more appropriate for long-term storage and applications necessitating meticulous regulation of DME release, such as in industrial chemical processes or sophisticated energy systems.

Chemical storage techniques for DME provide considerable safety benefits compared to conventional storage methods, as they mitigate the hazards linked to high-pressure or cryogenic storage. Nevertheless, they also present new challenges, including the necessity for meticulous regulation of reaction conditions and the possibility of byproduct generation. Environmental factors, including the energy demands for the synthesis and liberation of DME from chemical storage systems, must be considered to guarantee the sustainability of these methods. The advancement of sustainable catalysts and energy-efficient methods is essential for reducing the environmental impact of chemical storage.

Current research aims to optimize chemical storage techniques for DME to improve efficiency, safety, and scalability. Progress in materials science, including the creation of innovative catalysts and porous materials, may enhance the formation and release kinetics of DME hydrates and clathrates. Innovations in chemical engineering, such as the development of more efficient reaction systems and the incorporation of renewable energy sources, may improve the feasibility of reversible chemical storage techniques. Moreover, the advancement of intelligent storage systems featuring real-time monitoring and control functionalities could enhance the safety and reliability of chemical storage.

In summary, the chemical storage of DME presents a viable alternative to conventional storage techniques, offering improved safety, stability, and regulation of DME release. The utilization of hydrates and clathrates, along with reversible chemical reactions, offers distinct benefits and obstacles, rendering them appropriate for various applications. Ongoing research and innovation in this domain are crucial for surmounting current limitations and facilitating the extensive implementation of chemical storage techniques for DME.

6.2.3 Material-Based Storage

The material-based storage of DME exemplifies a sophisticated and adaptable method for handling this valuable chemical by utilizing the principles of adsorption and absorption. These techniques employ specialized materials to capture and retain DME, facilitating safe, efficient, and controllable storage and release. Material-based storage is especially beneficial because it functions at lower pressures and temperatures, thereby improving safety and facilitating handling relative to conventional storage methods. Material-based storage employs two primary mechanisms: adsorption and absorption, each utilizing different materials and methods to optimize DME storage.

6.2.3.1 Adsorption

Adsorption entails the attachment of DME molecules to the surfaces of porous materials, characterized by extensive surface areas and customized pore structures that enhance the capture and retention of gas molecules. Frequently utilized adsorbent materials encompass activated carbon, zeolites, and metal-organic frameworks (MOFs). These materials possess large surface areas, adjustable pore sizes, and chemical stability, rendering them highly efficient for DME storage [13].

Activated carbon is a prevalent adsorbent owing to its extensive surface area, economic viability, and accessibility. The porous structure facilitates the physical adsorption of DME molecules, which are retained by van der Waals forces. Activated carbon is especially appropriate for applications necessitating moderate storage capacities and straightforward regeneration [14].

Zeolites are crystalline aluminosilicates characterized by uniform pore dimensions and elevated thermal stability. Their clearly delineated structures facilitate selective adsorption of DME, rendering them optimal for applications necessitating meticulous regulation of storage and release. Zeolites are frequently employed in industrial processes where purity and efficiency are paramount [15].

Metal-organic frameworks (MOFs) are a category of sophisticated materials consisting of metal ions coordinated with organic ligands, resulting in highly porous structures with remarkable surface areas. Metal-organic frameworks (MOFs) provide exceptional adaptability in adjusting pore dimensions and chemical characteristics, facilitating enhanced DME adsorption capacities and release kinetics. Their substantial storage capacity and selectivity render MOFs a promising material for advanced

DME storage systems. Adsorption-based storage systems function at reduced pressures and temperatures relative to conventional methods, thereby mitigating the hazards linked to high-pressure or cryogenic storage. The release of DME can be regulated by modifying temperature or pressure conditions, offering a flexible and controllable storage solution [16].

6.2.3.2 Absorption

Absorption, unlike adsorption, entails the integration of DME into the interior of a material, where it is held within the material's structure rather than on its surface. This method employs materials like ionic liquids and particular polymers, which can absorb substantial amounts of DME and release it under regulated conditions.

Ionic liquids are salts that exist in a liquid form at ambient temperature, distinguished by their low volatility, high thermal stability, and adjustable chemical properties. They can assimilate DME via physical dissolution or chemical interactions, offering a stable and efficient storage medium. Ionic liquids are especially beneficial for applications that necessitate elevated storage capacities and regulated release mechanisms.

Specific polymers, including polyvinyl alcohol (PVA) and polyethylene glycol (PEG), can absorb DME owing to their flexible molecular configurations and attraction to small molecules. These polymers can be designed to enhance their absorption capacities and release kinetics, rendering them appropriate for various applications, including portable fuel systems and industrial processes [17].

Absorption-based storage systems provide the benefit of elevated storage densities and the capability to release DME via alterations in temperature, pressure, or chemical conditions. This renders them exceptionally versatile and adaptable to diverse storage and delivery needs.

The selection between adsorption and absorption techniques is contingent upon the particular demands of the application, encompassing storage capacity, release kinetics, and operational parameters. Adsorption is especially appropriate for applications necessitating swift absorption and desorption of DME, such as in fuel storage systems or gas separation processes. Absorption is more appropriate for applications necessitating elevated storage densities and regulated release, such as in chemical synthesis or energy storage systems.

Material-based storage techniques provide considerable safety benefits compared to conventional storage methods, as they function at reduced pressures and temperatures, thereby minimizing the risks of leaks, explosions, or environmental contamination. The creation of efficient and sustainable materials is essential for reducing the environmental impact of these methods. The synthesis of MOFs and ionic liquids frequently entails energy-intensive procedures and the utilization of rare or toxic substances, necessitating the advancement of more sustainable alternatives.

Current research aims to optimize material-based storage systems for DME to improve their efficiency, safety, and scalability. Progress in materials science, including the creation of innovative adsorbents and absorbents with enhanced capacities and accelerated kinetics, is essential for optimizing the performance of these systems. The incorporation of smart materials with real-time monitoring and control functionalities could further augment the safety and reliability of DME storage.

In conclusion, the material-based storage of DME, employing adsorption and absorption mechanisms, presents a viable alternative to conventional storage methods. These methodologies offer improved safety, efficiency, and controllability, rendering them appropriate for diverse applications. Ongoing research and innovation in this domain are crucial for surmounting current limitations and facilitating the extensive implementation of material-based storage systems for DME.

6.3 DME Transport

DME is a multifunctional compound employed as a substitute fuel, a propellant, and a raw material in chemical processes. Because of its characteristics, DME can be transported in different states, with its liquefied form under moderate pressure being the most prevalent. Transporting DME necessitates meticulous attention to safety protocols, infrastructure, and environmental factors to guarantee its secure and efficient conveyance. DME is predominantly conveyed in its liquefied state, enabling efficient storage and transportation at moderate pressures, usually ranging from 5 to 10 bar (70–145 psi) at normal temperatures.

6.3.1 Maritime Transport

Tankers specifically designed for the transportation of liquefied gases, such as liquefied petroleum gas (LPG), are used for long-distance bulk transportation across oceans in the case of DME. The tankers are fitted with cargo tanks that are both insulated and pressure-controlled, usually constructed from steel, in order to preserve the DME in its liquefied form. The tankers have a capacity that can reach tens of thousands of metric tons, which makes them well-suited for transporting DME on a large scale. Stringent safety protocols, such as the use of pressure relief valves, gas detection systems, and emergency venting, are employed to mitigate the hazards involved in transporting liquefied DME. The maritime transportation of DME is regulated by international standards, such as the International Maritime Organization's (IMO) International Code for the Construction and Equipment of Ships Carrying Liquefied Gases in Bulk (IGC Code). These regulations guarantee safety and adherence to the rules in various jurisdictions.

Inland waterways employ barges that are equipped with tanks that are pressurized and insulated for the purpose of transporting DME. Although barges have smaller capacities than ocean-going tankers, usually ranging from hundreds to thousands of metric tons, they provide a versatile and cost-efficient method for transporting

DME over shorter distances. Barges are especially advantageous in areas with highly developed river networks, as they can expedite the transportation of DME to locations that are not readily reachable by alternative modes of transport.

6.3.2 Land Transport

Tank trucks are commonly used to transport DME for localized distribution. These vehicles are equipped with pressurized tanks specifically designed to maintain DME in its liquid form while being transported. Tank trucks offer versatility in terms of delivery locations, enabling the transportation of DME to different areas, including remote regions that do not have pipeline infrastructure. While tank trucks have a lower capacity compared to rail tank cars or maritime vessels, typically in the range of thousands of liters, they play a crucial role in ensuring the final delivery of DME.

Rail tank cars are specifically designed for the efficient transportation of large quantities of DME over extensive land routes. The rail cars are made from durable materials that can withstand the necessary pressures to maintain the liquefied state of DME. They are also equipped with safety features like pressure relief valves and thermal insulation. Rail tank cars have the capacity to transport a significant volume of DME, amounting to tens of thousands of liters. This makes them a highly efficient and economically viable choice for distributing DME on a large scale, particularly in areas that have well-developed rail networks.

6.3.3 Pipelines

Pipelines provide a highly effective and eco-friendly method for the uninterrupted and extensive transportation of DME, especially when covering considerable distances. The pipelines are specifically engineered to function under moderate pressure in order to maintain the liquefied form of DME and are made from materials that possess the ability to withstand the corrosive properties of DME. Although the initial capital required for pipeline infrastructure is substantial, pipelines offer an economically efficient method for transporting significant quantities of DME while causing minimal environmental harm during operation. Currently, DME pipelines are only used in regions where there are large production and consumption centers. However, their significance is expected to increase as the demand for DME rises.

While not as prevalent, DME can be transported in its gaseous form for industrial purposes. This technique utilizes compressed gas cylinders or tube trailers specifically engineered to withstand the necessary pressures for storing DME in its gaseous state. Transporting gaseous DME necessitates specialized handling and adherence to strict safety protocols due to the flammability and potential health risks associated with DME in its gaseous state. Additionally, Table 6.1 provides a comparative overview of the different transportation modes for DME, highlighting their advantages, disadvantages, and typical capacities. When deciding how to transport DME, several factors need to be considered, such as the amount of DME to be transported, the distance it needs to travel, the presence of necessary infrastructure, and considerations regarding cost, safety, and environmental impact. For instance, tankers are well-suited for transporting large quantities over long distances by sea, whereas tank trucks provide versatility for delivering goods to specific locations. Pipelines, despite necessitating substantial initial capital, offer an effective and environmentally friendly solution for the continuous transportation of DME on a large scale.

With the increasing adoption of DME, especially as a substitute for fuel, it is imperative to make significant progress in transportation technologies to keep up with the rising demand. Conducting research on transportation methods that are more efficient and safer, as well as building infrastructure to facilitate the widespread distribution of DME, will be crucial for ensuring the sustainable expansion of the

Table 6.1 Overview of transportation modes for formic acid: advantages, disadvantages, and typical capacity

Mode of transportation	Advantages	Disadvantages	Typical capacity
Maritime (tankers)	High capacity, suitable for long distances, relatively low cost per unit	High initial investment, potential for marine pollution in case of spills, limited accessibility to inland locations	Tens of thousands of metric tons
Maritime (barges)	Cost-effective for inland waterways, flexible	Limited capacity, restricted to navigable waterways, slower than rail or pipeline	Hundreds to thousands of metric tons
Land (rail tank cars)	Efficient for bulk transport over land, well-established infrastructure, relatively safe	Limited flexibility, potential for accidents and spills, requires access to rail network	Tens of thousands of liters
Land (tank trucks)	Versatile, suitable for localized deliveries, flexible routing	Lower capacity, higher cost per unit compared to rail or pipeline	Thousands of liters
Land (pipelines)	Cost-effective for long distances, minimal environmental impact during operation, high capacity	High initial investment, potential for leaks and land disruption, requires significant infrastructure	Varies, dependent on pipeline diameter and pressure
Gaseous (cylinders/tube trailers)	Suitable for smaller volumes and specific applications	Lower capacity, higher cost per unit, requires specialized handling equipment and safety measures	Up to several thousand liters (cylinders) or tens of thousands of liters (tube trailers)

DME market. The future success of DME transportation will depend on effectively managing safety, cost, and environmental factors, as is the case with other energy carriers.

References

1. Peinado C, Liuzzi D, Sluijter SN, Skorikova G, Boon J, Guffanti S et al (2024) Review and perspective: next generation DME synthesis technologies for the energy transition. Chem Eng J 479. https://doi.org/10.1016/j.cej.2023.147494
2. Azizi Z, Rezaeimanesh M, Tohidian T, Rahimpour MR (2014) Dimethyl ether: a review of technologies and production challenges. Chem Eng Process Process Intensif 82:150–172. https://doi.org/10.1016/j.cep.2014.06.007
3. Turton R, Bailie RC, Whiting WB, Shaeiwitz JA, Bhattacharyya D (1998) Analysis, synthesis, and design of chemical processes. Choice Rev Online 36:36-0974–36-0974. https://doi.org/10.5860/CHOICE.36-0974
4. Al-Rabiah AA, Alshehri AS, Ibn Idriss A, Abdelaziz OY (2022) Comparative kinetic analysis and process optimization for the production of dimethyl ether via methanol dehydration over a γ-alumina catalyst. Chem Eng Technol 45:319–328. https://doi.org/10.1002/ceat.202100441
5. Wild S, Polierer S, Zevaco TA, Guse D, Kind M, Pitter S et al (2021) Direct DME synthesis on CZZ/H-FER from variable CO_2/CO syngas feeds. RSC Adv 11:2556–2564. https://doi.org/10.1039/d0ra09754c
6. Chang J, Fu Y, Luo Z (2012) Experimental study for dimethyl ether production from biomass gasification and simulation on dimethyl ether production. Biomass Bioenergy 39:67–72. https://doi.org/10.1016/j.biombioe.2011.01.044
7. Parvez AM, Wu T, Li S, Miles N, Mujtaba IM (2018) Bio-DME production based on conventional and CO_2-enhanced gasification of biomass: a comparative study on exergy and environmental impacts. Biomass Bioenergy 110:105–113. https://doi.org/10.1016/j.biombioe.2018.01.016
8. Parvez AM, Mujtaba IM, Hall P, Lester EH, Wu T (2016) Synthesis of bio-dimethyl ether based on carbon dioxide-enhanced gasification of biomass: process simulation using aspen plus. Energy Technol 4:526–535. https://doi.org/10.1002/ente.201500349
9. AFDC (2014) Alternative fuels data center, vol 51. https://doi.org/10.5860/choice.51-3266
10. Catizzone E, Freda C, Braccio G, Frusteri F, Bonura G (2021) Dimethyl ether as circular hydrogen carrier: catalytic aspects of hydrogenation/dehydrogenation steps. J Energy Chem 58:55–77. https://doi.org/10.1016/j.jechem.2020.09.040
11. Malla BK, Vishwakarma G, Chowdhury S, Nayak SK, Yamijala SSRKC, Pradeep T (2024) Formation and dissociation of dimethyl ether clathrate hydrate in interstellar ice mimics. J Phys Chem C 128:2463–2470. https://doi.org/10.1021/acs.jpcc.3c07792
12. Grzesik M, Skrzypek J (1999) Chemical equilibria in direct synthesis of dimethyl ether. Stud Surf Sci Catal 122:407–410. https://doi.org/10.1016/s0167-2991(99)80173-6
13. Tyraskis I, Capa A, Skorikova G, Sluijter SN, Boon J (2025) Performance optimization of sorption-enhanced DME synthesis (SEDMES) from captured CO_2 and renewable hydrogen. Front Chem Eng 7:1–13. https://doi.org/10.3389/fceng.2025.1521374
14. Palomo J, Rodríguez-Cano MÁ, Rodríguez-Mirasol J, Cordero T (2024) Biomass-derived activated carbon catalysts for the direct dimethyl ether synthesis from syngas. Fuel 365. https://doi.org/10.1016/j.fuel.2024.131264
15. Lotfollahzade Moghaddam A, Ghavipour M, Kopyscinski J, Hazlett MJ (2024) Methanol dehydration to dimethyl ether over KFI zeolites. Effect of template concentration and crystallization time on catalyst properties and activity. Appl Catal A Gen 672. https://doi.org/10.1016/j.apcata.2024.119594

16. Wright KR, Nath K, Matzger AJ (2022) Superior metal-organic framework activation with dimethyl ether. Angew Chemie 134. https://doi.org/10.1002/ange.202213190
17. Powers MN, Rice TE, Chowdhury A, Mansha MW, Hella MM, Wilke I et al (2023) Dimethyl ether gas sensing using rotational absorption spectroscopy in the THz frequency region from 220 to 330 GHz. Sensors Actuat B Chem 384. https://doi.org/10.1016/j.snb.2023.133635

Open Access This chapter is licensed under the terms of the Creative Commons Attribution 4.0 International License (http://creativecommons.org/licenses/by/4.0/), which permits use, sharing, adaptation, distribution and reproduction in any medium or format, as long as you give appropriate credit to the original author(s) and the source, provide a link to the Creative Commons license and indicate if changes were made.

The images or other third party material in this chapter are included in the chapter's Creative Commons license, unless indicated otherwise in a credit line to the material. If material is not included in the chapter's Creative Commons license and your intended use is not permitted by statutory regulation or exceeds the permitted use, you will need to obtain permission directly from the copyright holder.

Chapter 7
Formic Acid

7.1 Formic Acid Production

Formic acid derives its name from the Latin word for ants (Formica), discovered upon the observations and experiments made with an Acid Juyce found in ants in the early 70 s. With the chemical formula (HCOOH), formic acid, the simplest yet strongest organic acid, is more eco-friendly and biodegradable than other organic and inorganic acids and widely applied in agricultures, rubbery, pharmaceuticals, animal feeds, leather, and textiles industries. Formic acid is not only a valuable chemical commodity but also a crucial energy carrier and medium, offering potential solutions to the energy crisis and supporting the development of renewable energy infrastructures [1]. It can be directly utilized or upgraded into a range of high-value fuels, including hydrogen (H_2), carbon monoxide (CO), methanol, and bio-oils, enhancing its role in the transition to sustainable energy systems.

In the chemical industry, formic acid is produced by four main processes: preparation of free formic acid from formate salts, oxidation of hydrocarbons (incl., biomass), hydrolysis of formamide, and methyl formate hydrolysis.

7.1.1 Methyl Formate Hydrolysis

The production of formic acid by hydrolysis of methyl formate ($HCOOCH_3$) has two key stages: carbonylation of methanol and hydrolysis of the methyl formate produced. Methanol generated in this process is recycled back to the initial stage, promoting efficiency and minimizing waste. The two stages involved are represented with chemical equations and their heat of reactions.

$$CH_3OH + CO \rightarrow HCOOCH_3 \quad \Delta H^\circ = -29 \frac{kJ}{mol}$$

$$\text{HCOOCH}_3 + \text{H}_2\text{O} \rightleftharpoons \text{CH}_3\text{OH} + \text{HCOOH} \quad \Delta H° = 16.3 \frac{\text{kJ}}{\text{mol}}$$

The carbonylation of methanol, catalyzed by basic agents such as sodium or potassium methoxide, is a well-established and reliable industrial process. However, the hydrolysis of methyl formate presents significant technological challenges. The hydrolysis is a reversible reaction with a low equilibrium constant, making the process inherently unfavorable. The equilibrium can shift based on water concentration, necessitating large excesses of water to drive the reaction forward. This, in turn, introduces the challenge of efficiently removing the excess water without excessive energy consumption. Additionally, methyl formate's high volatility (boiling point: 31.5 °C) complicates the process, as any residual methyl formate can readily vaporize. Formic acid, being a strong acid, can further catalyze the re-esterification of methyl formate, making it difficult to separate unreacted methyl formate without inducing significant re-esterification. This interplay of volatility, reversibility, and reactivity underscores the complexity of formic acid production through methyl formate hydrolysis.

Despite the complexities discussed, the production of formic acid via methyl formate hydrolysis is not only feasible but actively implemented and continually evolving within the industry. Sharma et al. provide a detailed process configuration that outlines the production pathway [2]. In this process, fresh carbon monoxide (CO) is fed into a continuously stirred tank reactor (CSTR) at a rate of 70.3 kmol/h. To maintain an excess of methanol, makeup methanol is also introduced into the CSTR. The reactor operates under controlled conditions at 75 °C and 40.53 bar. The CSTR output undergoes flashing into a phase separator, with the resulting stream directed to a distillation column (C1) for the separation of methyl formate (HCOOCH_3) from methanol. The purified methyl formate is subsequently fed into a reactive distillation (RD) unit for further processing. A partial vapor–liquid condenser within the separation column purges residual CO, limiting total CO emissions to approximately 90 kg/h across all purge streams. Unconverted methanol collected from the bottom of the distillation column (C1), is recycled back to the CSTR. Formic acid and methanol are produced via the hydrolysis of methyl formate within the RD column, where separation occurs concurrently.

Fresh water, essential for the hydrolysis reaction, is introduced at the second tray of the RD column (i.e., with trays numbered from top to bottom). Meanwhile, methyl formate from the distillation unit (C1) enters at the 33rd tray, optimizing contact and reaction efficiency. This process configuration highlights the integration of reaction and separation units, underscoring advancements in reducing emissions and maximizing resource recovery in formic acid production.

7.1.2 Biomass Oxidation

Producing valuable products from abundant renewable resources, such as biomass, offers a promising pathway to mitigate climate change while providing manufacturers with profitable opportunities. Supercritical water (operating at 400–600 °C and 22–40 MPa), is frequently used as a green solvent for biomass transformations particularly in an acid- and base-catalyzed reactions (i.e., generates protons and hydroxides). Several studies, in early 2000s, reported utilization of supercritical water for biomass conversion producing organic acids: formic acid, acetic acid, lactic acid and others after glucose conversion (i.e., plant-based biomass), in which air acted as the oxidant. By designing a catalyst for selective formic acid production from biomass, thereby, a large-scale implementation would be upon reach. However, ambitions were quickly met with challenges. Small to below than average yields of 15–38% max under strong oxidants and high operating conditions, causing severe decomposition of the liquid organic acids into gaseous products (e.g., CO_2) were recorded.

A fundamentally different approach to produce formic acid from biomass is suggested by Reicher et al. in 2015 [3], namely OxFa. With a very broad range of applicable biomasses (e.g., mono- and disaccharides as well as water-insoluble substrates like cellulose, and spruce and beech wood), the experiment employs [$H_5PV_2Mo_{10}O_{40}$] (HPA) as a catalyst to convert a full substrate to FA (formic acid) as the sole product in a liquid phase accompanied by CO_2 as gaseous by-product. At mild conditions of 89 °C and 20 bar, a complete conversion of glucose to 49% FA and 51% CO_2 after 3 h of reaction time. The highest reported yield of formic acid recorded upon experimenting with process parameters: reaction conditions, media, catalyst, biomaterial, etc., is 61% for a selectivity of 68% [3]. This process avoids many problems related to biorefinery reactions including, complex product mixtures, sticky polymer or tar formation in unit operations [3].

7.1.3 Electrochemical Reduction of CO_2

The electrochemical reduction of CO_2 into value-added chemicals, (e.g., formic, and acetic acids), holds immense promise for mitigating carbon emissions while producing valuable goods. By merging renewable energy means to power the electrochemical reduction of CO_2 with CCU, the switch from fossil fuels seem to be virtually plausible.

At the cathode, the electrochemical reduction of CO_2 could, however, follow different pathways involving two, four, six, and eight-electron reductions listed as follows.

$$CO_{2(g)} + 2H^+ + 2e^- \rightarrow HCOOH$$
$$CO_{2(g)} + 2H_2O_{(l)} + 2e^- \rightarrow HCOO_{(aq)} + OH^-$$

$$CO_{2(g)} + 2H^+ + 2e^- \rightarrow CO_{(g)} + H_2O_{(l)}$$
$$CO_{2(g)} + 2H_2O_{(l)} + 2e^- \rightarrow CO_{(g)} + 2OH^-$$
$$CO_{2(g)} + 4H^+ + 4e^- \rightarrow CH_2O_{(l)} + H_2O_{(l)}$$
$$CO_{2(g)} + 3H_2O_{(l)} + 4e^- \rightarrow CH_2O_{(l)} + 4OH^-$$
$$CO_{2(g)} + 6H^+ + 6e^- \rightarrow CH_3OH_{(l)} + H_2O_{(l)}$$
$$CO_{2(g)} + 5H_2O_{(l)} + 6e^- \rightarrow CH_3OH_{(l)} + 6OH^-$$
$$CO_{2(g)} + 8H^+ + 8e^- \rightarrow CH^+_{4(g)} + 2H_2O_{(l)}$$
$$CO_{2(g)} + 6H_2O_{(l)} + 8e^- \rightarrow CH^+_{4(g)} + 8OH^-$$

At the anode, water oxidation (electrolysis):

$$2H_2O \rightarrow 4H^+ + O_2 + 4e^-.$$

Depending on the choice of catalyst, one-carbon (C1) products can be fabricated as shown in the chemical equations. Metal-based catalyst scored high selectivity and Faradaic efficiency for the production of C1 products in which tin and tin-alloys are found to be the most efficient for formic acid (88% selectivity) [4].

For the formation of HCOOH, a commonly accepted method involves the electronation of CO_2, via the addition of an electron, resulting in the formation of radical carbon dioxide (CO_2^-) overcoming the thermodynamic potential of -1.90 V versus normal hydrogen electrode (NHE). As the reactions progress, adsorbed HCO_2 undergoes further electronation and generates ($HCOO^-$) ions, while the successive reactions of CO_2^- and H_2O produce hydroxide ions (OH^-). The cathodic reaction for the electroreduction of CO_2 to form formic acid is as follows

$$CO_{2(g)} + 4H^+ + 4e^- \rightarrow 2HCOOH.$$

The overall redox reaction:

$$2H_2O + CO_{2(g)} \rightarrow 2HCOOH + O_2.$$

Reports suggest that formic acid/formate formation occurs through four potential pathways, although the details of each remain subject of ongoing research.

1. **Pathway 1:** Weak binding of the CO_2 anion radical to the catalyst leads to HCOOH via hydration. Protonation occurs as the intermediate binds to the metal electrode through one or two oxygen atoms.
2. **Pathway 2:** Metal (M)-C bonding forms the $HCOO^-$ or *COOH intermediate. Protonation and reduction yield HCOOH, or possibly CO and H_2O.
3. **Pathway 3:** M–O bonds generate the $*$ OCHO intermediate, where CO_2 binds through O_2, leading to formic acid formation.
4. **Pathway 4:** CO_2 inserts into the M–H bond, producing $*$ OCHO, followed by reduction to HCOOH.

7.1 Formic Acid Production

At tin oxide electrodes, SnO_2 reduces to SnII oxyhydroxide, which reacts with CO_2 to form surface-bound carbonate, eventually converting to formate through proton and electron transfer.

7.1.4 Hydrolysis of Formamide

Formic acid production via formamide hydrolysis involves the reaction of formamide ($HCONH_2$) with water to yield formic acid (HCOOH) and ammonia (NH_3). This reaction can be represented as follows:

$$HCONH_2 + H_2O \rightarrow HCOOH + NH_3.$$

The hydrolysis of formamide to produce formic acid offers a simple, cost-effective, and potentially sustainable route for industrial applications. Even the reagent (i.e., formamide) used for formic acid production, can be synthesized from using readily available and inexpensive feedstocks: CO_2 and ammonia (NH_3), as proposed by Shao et al., through electrosynthesis [5].

$$CO_2 + NH_3 \rightarrow HCONH_2$$

Therefore, the process can contribute to carbon recycling and align with green chemistry principles. Additionally, the reaction avoids hazardous reagents, reducing environmental impact and enhancing safety. The ammonia byproduct can be recovered and reused, further improving resource efficiency. As formic acid plays a key role in agriculture, textiles, and hydrogen storage, formamide hydrolysis presents a scalable and environmentally friendly alternative for meeting growing industrial demand.

According to [6], this process typically occurs under elevated temperatures (100–200 °C) and can be catalyzed by either acid or base catalysts, such as sulfuric acid or sodium hydroxide, to accelerate the reaction. The hydrolysis reaction breaks the C–N bond in formamide, yielding formic acid and ammonia, which can be separated and purified by distillation. The ammonia byproduct can be recovered and reused in other applications, such as fertilizer production. The simplicity and efficiency of this process, combined with its potential for using CO_2 as a feedstock for formamide, make it a promising route for producing formic acid in an environmentally sustainable manner.

7.2 Formic Acid Storage

Formic acid, a carboxylic acid, is widely employed in diverse industries such as agriculture, leather production, and as a preservative. Given its corrosive and volatile properties, the storage of formic acid necessitates meticulous deliberation. Formic acid can be stored using three main methods: physical storage, chemical storage, and material-based storage. Figure 7.1 categorizes formic acid storage into three primary methods: physical storage, chemical storage, and material-based storage. Physical storage utilizes stainless steel tanks, high-density polyethylene (HDPE) tanks, and glass containers for laboratory applications to prevent corrosion and leakage. Chemical storage includes conversion into salts such as sodium formate and calcium formate, as well as formic acid esters (e.g., methyl formate, ethyl formate) for controlled release and stability. Material-based storage involves adsorption and absorption techniques, including activated carbon, silica gel, polymers, and hydrogels, providing enhanced containment and controlled release mechanisms. These storage methods are essential for safe handling, transportation, and industrial utilization of formic acid in various applications, including energy storage and chemical synthesis.

7.2.1 Physical Storage

Formic acid, a highly corrosive and reactive chemical, necessitates specialized storage solutions to maintain its stability and safety. Owing to its corrosive properties, formic acid can deteriorate various materials, rendering the choice of suitable storage containers essential. The predominant materials utilized for formic acid storage are stainless steel and high-density polyethylene (HDPE), both demonstrating superior resistance to corrosion and chemical degradation. These materials are extensively utilized in industrial and commercial environments for the safe and efficient storage of formic acid.

7.2.1.1 Stainless Steel

Stainless steel is the material of choice for formic acid storage because of its superior corrosion resistance and mechanical strength. Stainless steel tanks are commonly employed for extensive storage in industrial settings, as they can endure acidic conditions and preserve the integrity of stored formic acid over prolonged durations. The utilization of stainless steel guarantees durability and longevity, minimizing the necessity for frequent replacements or repairs [7].

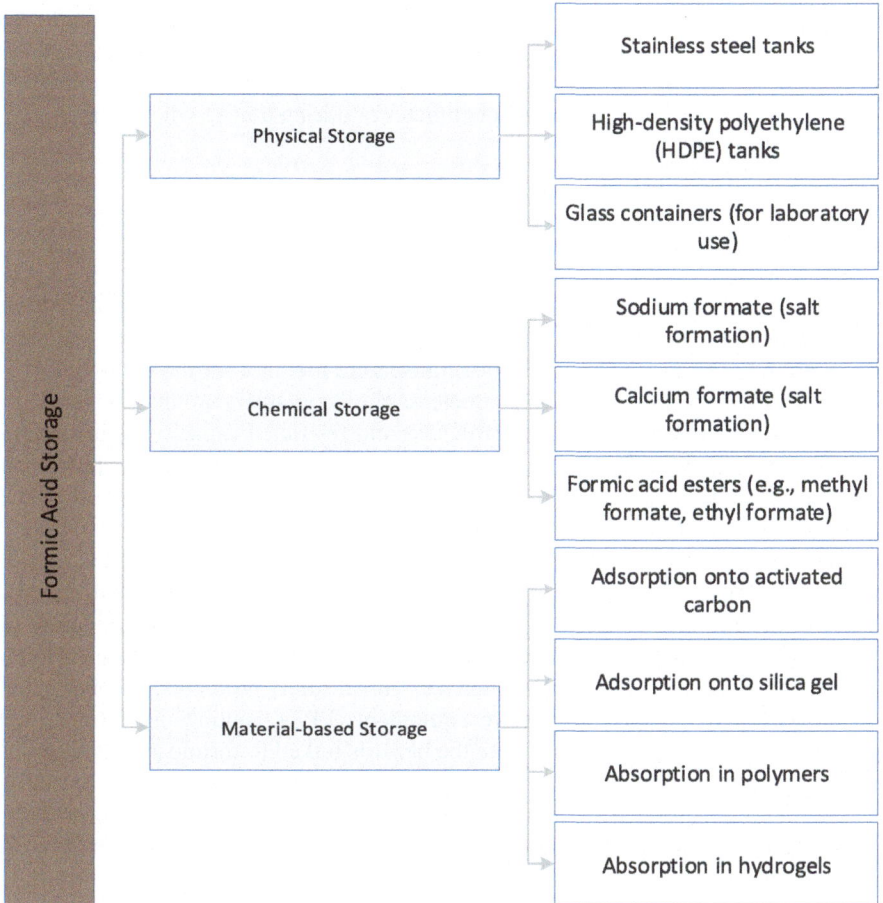

Fig. 7.1 Classification of formic acid storage methods

7.2.1.2 High-Density Polyethylene

High-density polyethylene (HDPE) is a commonly utilized material for the storage of formic acid, especially in smaller-scale applications or situations necessitating portability. HDPE containers are lightweight, economical, and exceptionally resistant to chemical corrosion, rendering them appropriate for the transportation and storage of formic acid in diverse environments. Both stainless steel and HDPE containers are engineered to avert leaks and reduce the likelihood of exposure to the corrosive acid [8].

Effective ventilation and temperature regulation are critical elements of formic acid storage systems. Formic acid can emit vapors that are detrimental if inhaled, and the buildup of these vapors in enclosed areas can present considerable safety hazards. Storage tanks are consequently outfitted with ventilation systems to dissipate vapors and ensure a safe environment. Moreover, temperature regulation systems are essential to avert overheating, which can hasten the degradation of formic acid and result in the emission of hazardous gases like carbon monoxide. Ensuring a consistent temperature is especially crucial in hot climates or during transit, where temperature variations may arise.

7.2.1.3 Glass Containers

Formic acid is typically stored in glass containers, such as borosilicate glass bottles, for laboratory-scale applications. Glass exhibits significant resistance to chemical degradation by formic acid and offers a transparent medium for the facile observation of the contained substance. Nonetheless, glass containers are inappropriate for extensive storage because of their brittleness and restricted capacity. In laboratory environments, glass containers must be tightly sealed to avert leakage and reduce air exposure, which can lead to oxidation and decomposition of the acid [9].

Securing the airtight sealing of storage containers is an essential safety protocol to avert leaks and mitigate the risk of exposure to formic acid. Leakage may result in chemical burns, environmental pollution, and the emission of hazardous vapors. Effective sealing also reduces exposure to air, which can lead to the oxidation of formic acid, resulting in its decomposition into carbon dioxide and water. Alongside secure sealing, storage facilities must establish safety protocols, encompassing the utilization of personal protective equipment (PPE), routine inspections, and emergency response plans, to alleviate the hazards linked to formic acid storage.

Current research and innovation aim to enhance the safety and efficiency of formic acid storage. Progress in materials science, including the creation of corrosion-resistant coatings and composite materials, may improve the durability and performance of storage containers. Furthermore, the incorporation of intelligent monitoring systems, such as sensors for leak detection or temperature fluctuations, could enhance the safety and reliability of formic acid storage facilities.

In conclusion, formic acid storage necessitates specialized containers constructed from materials like stainless steel or HDPE, in addition to sufficient ventilation and temperature regulation systems to guarantee safety and stability. Glass containers are appropriate for laboratory-scale storage but impractical for large-scale applications. Ensuring secure sealing and compliance with regulatory standards is crucial to avert leaks and reduce the risks linked to formic acid storage. Ongoing progress in materials and safety technologies will be essential for improving the efficiency and sustainability of formic acid storage systems.

7.2.2 Chemical Storage

The chemical storage of formic acid exemplifies a sophisticated method for handling this highly reactive and corrosive substance by stabilizing it via chemical reactions or integrating it into other compounds. This method improves the safety and stability of formic acid while offering a more convenient and regulated approach to storage and transportation. Two principal strategies are utilized in the chemical storage of formic acid: the generation of formates and the production of formic acid esters. Each method presents distinct benefits and obstacles, rendering them appropriate for particular applications based on variables such as storage duration, release mechanisms, and handling specifications.

7.2.2.1 Formation of Formates

An efficient approach for the chemical storage of formic acid entails its transformation into formates, which are salts produced by the reaction of formic acid with bases or metal oxides. Formates, including sodium formate (HCOONa) and calcium formate ($Ca(HCOO)_2$), are solid compounds that demonstrate enhanced stability and improved handling compared to pure formic acid. These salts exhibit non-volatility, non-corrosiveness, and reduced hazard potential, rendering them suitable for prolonged storage and transport [10].

Sodium formate is a commonly utilized formate owing to its high solubility in water and straightforward synthesis. It is frequently utilized in industrial applications, including de-icing agents, oil drilling fluids, and as a reducing agent in chemical processes. Sodium formate can be readily transformed back into formic acid via acidification, offering a dependable and regulated storage method. Calcium formate is a stable formate esteemed for its low solubility and high thermal stability. It is utilized in applications including animal feed additives, concrete additives, and as a preservative in the food industry. Similar to sodium formate, calcium formate can be reconstituted into formic acid via chemical reactions, thereby confirming its efficacy as a storage medium [11].

The utilization of formats for chemical storage presents numerous benefits, such as diminished volatility, enhanced safety, and streamlined handling. The regeneration of formic acid from formates necessitates supplementary chemical processing, potentially elevating operational complexity and expenses.

7.2.2.2 Formic Acid Esters

An alternative method for the chemical storage of formic acid entails its transformation into formic acid esters, organic compounds synthesized through the reaction of formic acid with alcohols. Formic acid esters, including methyl formate ($HCOOCH_3$) and ethyl formate ($HCOOCH_2CH_3$), demonstrate reduced volatility and enhanced stability relative to pure formic acid. These esters exist as liquids at ambient temperature, allowing for safer and more convenient storage and transportation [12].

Formic acid esters can be hydrolyzed into formic acid and the respective alcohol as required, facilitating a controlled release mechanism. This renders them especially advantageous for applications necessitating accurate dosing or controlled release of formic acid, such as in chemical synthesis or industrial operations. The hydrolysis reaction can be facilitated by acids or bases, enabling the effective regeneration of formic acid under regulated conditions.

The utilization of formic acid esters for chemical storage presents numerous advantages, such as diminished handling hazards, enhanced storage stability, and the capacity to regulate the release of formic acid. The synthesis of esters necessitates meticulous optimization to achieve elevated yields and purity, while the hydrolysis process must be meticulously controlled to prevent the generation of undesirable byproducts.

The selection between formats and formic acid esters for chemical storage is contingent upon the particular demands of the application. Formats are especially appropriate for long-term storage and applications necessitating stable, non-volatile substances, such as in industrial or agricultural contexts. Conversely, formic acid esters are more appropriate for applications necessitating liquid storage and regulated release, such as in chemical production or laboratory investigations.

Chemical storage techniques for formic acid provide considerable safety benefits compared to conventional storage methods, as they mitigate the hazards linked to the corrosive and volatile characteristics of pure formic acid. The regeneration of formic acid from formates or esters necessitates meticulous handling and processing to guarantee safety and efficiency. Environmental factors, including the energy demands of chemical reactions and the possibility of byproduct generation, must be considered to guarantee the sustainability of these methods.

Current research aims to optimize chemical storage techniques for formic acid to improve efficiency, safety, and scalability. Progress in chemical engineering, including the creation of more efficient catalysts and reaction systems, may enhance the synthesis and regeneration processes for formates and esters. The incorporation of renewable energy sources and sustainable practices into chemical storage systems will be crucial for mitigating the environmental impact of formic acid storage.

In conclusion, the chemical storage of formic acid via the synthesis of formates and esters presents a viable alternative to conventional storage techniques. These methodologies offer improved safety, stability, and controllability, rendering them appropriate for diverse applications. Ongoing research and innovation in this domain are crucial for surmounting current limitations and facilitating the extensive implementation of chemical storage techniques for formic acid.

7.2.3 Material-Based Storage

Material-based storage techniques provide an advanced and efficient means of optimizing formic acid storage by utilizing materials that can adsorb or absorb the acid. These methods enhance the safety and stability of formic acid storage while offering mechanisms for controlled release and diminished volatility, rendering them especially advantageous in industrial and laboratory environments. Material-based storage systems can effectively mitigate the hazards of formic acid, including its corrosiveness, volatility, and potential for decomposition, by employing materials such as activated carbon, silica gel, polymers, and gels.

7.2.3.1 Adsorption

Adsorption entails the capture and retention of formic acid molecules on the surfaces of porous materials characterized by high surface areas and customized pore structures. This process is governed by physical forces, including van der Waals interactions, or chemical interactions, such as hydrogen bonding. Activated carbon and silica gel are frequently utilized adsorbent materials, renowned for their substantial adsorption capacities and chemical stability.

Activated carbon is a highly porous substance with a vast surface area, rendering it an exceptional adsorbent for formic acid. The capacity to adsorb and retain formic acid molecules on its surface diminishes the likelihood of vapor emission and mitigates the dangers linked to liquid formic acid. Activated carbon is especially appropriate for applications that necessitate moderate storage capacities and straightforward regeneration, such as in industrial gas purification systems or emergency spill containment.

Silica gel is a highly efficient adsorbent, recognized for its consistent pore architecture and exceptional thermal stability. It can adsorb formic acid via physical interactions, offering a stable and regulated storage medium. Silica gel is frequently utilized in applications necessitating meticulous regulation of humidity and gas composition, such as in chemical production or laboratory settings. Adsorption-based storage systems function at ambient temperatures and pressures, thereby minimizing energy demands and safety hazards linked to high-pressure or cryogenic storage. The release of formic acid can be regulated by modifying environmental conditions, such as temperature or pressure, offering a flexible and controllable storage option.

7.2.3.2 Absorption

Absorption, unlike adsorption, entails the integration of formic acid into the interior of a material, where it is preserved within the material's matrix rather than on its exterior. This method employs materials like polymers and gels that can absorb substantial amounts of formic acid and release it under regulated conditions.

Absorption-based storage systems offer high storage densities and the capability to release formic acid in a controlled manner, rendering them suitable for applications necessitating precise dosing or gradual release. Nonetheless, the advancement of effective and stable absorbent materials continues to be a significant challenge, especially regarding the preservation of their performance across numerous absorption–desorption cycles.

Material-based storage techniques are especially beneficial in applications necessitating controlled release and diminished volatility of formic acid. In industrial processes, such as chemical synthesis or textile manufacturing, these methods can improve safety and efficiency by reducing the hazards linked to the handling of liquid formic acid. In regulated settings, such as laboratories or medical institutions, material-based storage systems can ensure accurate dosing and minimize the risk of unintentional exposure.

Material-based storage techniques provide substantial safety benefits compared to conventional storage methods by mitigating the hazards linked to the corrosive and volatile characteristics of formic acid. The advancement of sustainable and eco-friendly materials is essential for reducing the environmental impact of these methods. The synthesis of adsorbent and absorbent materials frequently entails energy-intensive procedures and the utilization of rare or hazardous chemicals, necessitating the advancement of more sustainable alternatives.

Current research aims to optimize material-based storage systems for formic acid to improve their efficiency, safety, and scalability. Progress in materials science, including the creation of innovative adsorbents and absorbents with enhanced capacities and accelerated kinetics, is essential for optimizing the performance of these systems. The incorporation of smart materials with real-time monitoring and control functionalities could further improve the safety and reliability of formic acid storage.

In conclusion, material-based storage techniques, employing adsorption and absorption mechanisms, present a promising strategy for optimizing formic acid storage. These methods offer improved safety, stability, and controllability, rendering them appropriate for various applications. Ongoing research and innovation in this domain are crucial for surmounting current limitations and facilitating the extensive implementation of material-based storage systems for formic acid.

7.3 Formic Acid Transport

Formic acid, a vital chemical utilized in diverse industries ranging from agriculture to energy storage, poses distinctive difficulties in transportation due to its corrosive nature and potential health risks. Ensuring the widespread use of transportation methods that are both efficient and safe is crucial for minimizing risks. Formic acid is primarily transported in its liquid state, with the mode of transportation selected based on the volume and distance of transport as follows:

7.3.1 Maritime Transport

Chemical tankers are specifically used for transporting large quantities of chemicals over long distances across oceans. The tankers are outfitted with cargo tanks that are resistant to corrosion, usually constructed from stainless steel or carbon steel coated with epoxy. This is done to prevent deterioration caused by the extremely corrosive properties of formic acid. Formic acid tankers are specifically engineered to transport massive quantities of metric tons per trip, rendering them a highly effective option for transporting on a large scale. In order to reduce the risks involved in handling formic acid, these tankers are equipped with sophisticated safety measures like vapor control systems and leak detection mechanisms, which promptly identify and handle any potential breaches. The maritime transportation of formic acid is subject to strict regulations aimed at preventing environmental contamination. These regulations are enforced through international maritime safety standards that dictate the design and operation of the vessels involved.

Barges equipped with tanks that are resistant to corrosion are used for transportation on inland waterways. Although barges have a smaller capacity than ocean-going tankers, typically handling hundreds to thousands of metric tons, they provide a flexible and cost-effective method for transporting formic acid over shorter distances. Barges are especially beneficial in areas with extensive river networks, as they offer an efficient mode of transportation to locations that are not reachable by road or rail.

7.3.2 Land Transport

Tank trucks provide a flexible method for distributing formic acid in a specific area. These trucks are equipped with tank trailers that are resistant to corrosion and are mounted on their chassis. This enables the secure transportation of formic acid in smaller quantities compared to rail tank cars. Tank trucks are highly suitable for transporting formic acid to various locations, even those that may be challenging to reach using alternative transportation methods, due to their flexibility. Although tank trucks have a smaller capacity, usually ranging in the thousands of liters, they play a vital role in connecting large production facilities with end-users or smaller storage sites.

Rail tank cars are the preferred choice for transporting large quantities of formic acid over long distances due to their capacity for bulk land transportation. These rail cars are made from materials that are compatible with formic acid, such as stainless steel. They are equipped with safety mechanisms, including pressure relief valves and leak detection systems, to guarantee the safe transportation of this corrosive chemical. Rail tank cars are commonly used to transport large quantities of formic acid, typically in the tens of thousands of liters. This method is both cost-effective and efficient, especially in areas with well-developed rail networks.

7.3.3 Pipelines

While the corrosive properties of formic acid make it less commonly used, dedicated pipelines can still be utilized for large-scale and continuous transportation. This is particularly applicable in regions near production facilities or major consumption hubs. These pipelines must be made from specialized materials that can withstand the corrosive properties of formic acid, and they need strict maintenance protocols to guarantee safe operation. Although pipelines provide substantial cost-effectiveness and have a positive environmental impact for long-distance transportation, their extensive adoption is hindered by the substantial initial investment required and the difficulties posed by the corrosive nature of formic acid.

Gaseous formic acid is sometimes used for specific purposes during transportation. This approach generally entails the utilization of pressurized gas cylinders or tube trailers that are specifically engineered to safely manage the potential risks associated with gaseous formic acid. Although less prevalent than liquid transportation, this approach is utilized for specific industrial processes that necessitate gaseous formic acid. Specialized handling and strict safety measures are necessary for the transportation of gaseous formic acid due to its potential health and safety risks. Table. 7.1 provides a comparative overview of the different transportation modes for formic acid, highlighting their advantages and disadvantages. Additionally, when deciding on the best way to transport formic acid, several factors need to be taken into account. These include the amount of formic acid to be transported, the distance it needs to travel, the presence of suitable infrastructure, and considerations regarding cost, safety, and environmental impact. Maritime tankers are well-suited for transporting large quantities over long distances, whereas tank trucks provide versatility for delivering goods within a specific area. Pipelines, while not as prevalent, offer an effective solution for the uninterrupted, extensive transportation in particular circumstances.

The increasing global demand for formic acid necessitates the development of innovative and sustainable transportation solutions. Conducting research on safer and

Table. 7.1 Overview of transportation modes for formic acid: advantages and disadvantages

Mode of transportation	Advantages	Disadvantages
Maritime (tankers)	High capacity, suitable for long distances, relatively low cost per unit	High initial investment, potential for marine pollution in case of spills, limited accessibility to inland locations
Maritime (barges)	Cost-effective for inland waterways, flexible	Limited capacity, restricted to navigable waterways, slower than rail or pipeline
Land (rail tank cars)	Efficient for bulk transport over land, well-established infrastructure, relatively safe	Limited flexibility, potential for accidents and spills, requires access to rail network
Land (tank trucks)	Versatile, suitable for localized deliveries, flexible routing	Lower capacity, higher cost per unit compared to rail or pipeline
Land (pipelines)	Cost-effective for long distances, minimal environmental impact during operation, high capacity	High initial investment, potential for leaks and land disruption, less common for formic acid due to its corrosive nature
Gaseous (cylinders/tube trailers)	Suitable for smaller volumes and specific applications	Lower capacity, higher cost per unit, requires specialized handling equipment and safety measures

more efficient transportation methods, as well as making advancements in handling and storage technologies, will play a vital role in supporting the growing uses of formic acid. Prioritizing the safety of individuals and the environment will continue to be of utmost importance as the industry adjusts to meet increasing demands.

References

1. Supronowicz W, Ignatyev IA, Lolli G, Wolf A, Zhao L, Mleczko L (2015) Formic acid: a future bridge between the power and chemical industries. Green Chem 17:2904–2911. https://doi.org/10.1039/c5gc00249d
2. Sharma S, Patle DS, Gadhamsetti AP, Pandit S, Manca D, Nirmala GS (2018) Intensification and performance assessment of the formic acid production process through a dividing wall reactive distillation column with vapor recompression. Chem Eng Process Process Intensif 123:204–13. https://doi.org/10.1016/j.cep.2017.11.016
3. Reichert J, Brunner B, Jess A, Wasserscheid P, Albert J (2015) Biomass oxidation to formic acid in aqueous media using polyoxometalate catalysts—boosting FA selectivity by in-situ extraction. Energy Environ Sci 8:2985–2990. https://doi.org/10.1039/c5ee01706h
4. Orlić M, Hochenauer C, Nagpal R, Subotić V (2024) Electrochemical reduction of CO_2: a roadmap to formic and acetic acid synthesis for efficient hydrogen storage. Energy Convers Manag 314. https://doi.org/10.1016/j.enconman.2024.118601
5. Shao J, Meng N, Wang Y, Zhang B, Yang K, Liu C et al (2022) Scalable electrosynthesis of formamide through C−N coupling at the industrially relevant current density of 120 mA cm^{-2}. Angew Chemie 134. https://doi.org/10.1002/ange.202213009
6. Gorb L, Asensio A, Tuñón I, Ruiz-López MF (2005) The mechanism of formamide hydrolysis in water from Ab initio calculations and simulations. Chem A Eur J 11:6743–6753. https://doi.org/10.1002/chem.200500346
7. Sulejmanovic D, Jun J, Keiser J (2022) Effect of formic and acetic acids on corrosion of 410 stainless steel in bio-oils. ECS Meet Abstr MA2022-02:692–692. https://doi.org/10.1149/ma2022-0210692mtgabs
8. Álvarez A, Bansode A, Urakawa A, Bavykina AV, Wezendonk TA, Makkee M et al (2017) Challenges in the greener production of formates/formic acid, methanol, and DME by heterogeneously catalyzed CO_2 hydrogenation processes. Chem Rev 117:9804–9838. https://doi.org/10.1021/acs.chemrev.6b00816
9. Bacquart T, Morris ASO, Hookham M, Ward MKM, Underwood R, Hristova Y et al (2023) Formic acid in hydrogen: is it stable in a gas container? Processes 11. https://doi.org/10.3390/pr11061748
10. Fernández-Caso K, Díaz-Sainz G, Alvarez-Guerra M, Irabien A (2023) Electroreduction of CO_2: advances in the continuous production of formic acid and formate. ACS Energy Lett 8:1992–2024. https://doi.org/10.1021/acsenergylett.3c00489
11. Xu F, Huang W, Wang Y, Astruc D, Liu X (2022) Efficient and controlled H_2 release from sodium formate. Inorg Chem Front 9:3514–3521. https://doi.org/10.1039/d2qi00774f
12. Mellmann D, Sponholz P, Junge H, Beller M (2016) Formic acid as a hydrogen storage material—development of homogeneous catalysts for selective hydrogen release. Chem Soc Rev 45:3954–3988

Open Access This chapter is licensed under the terms of the Creative Commons Attribution 4.0 International License (http://creativecommons.org/licenses/by/4.0/), which permits use, sharing, adaptation, distribution and reproduction in any medium or format, as long as you give appropriate credit to the original author(s) and the source, provide a link to the Creative Commons license and indicate if changes were made.

The images or other third party material in this chapter are included in the chapter's Creative Commons license, unless indicated otherwise in a credit line to the material. If material is not included in the chapter's Creative Commons license and your intended use is not permitted by statutory regulation or exceeds the permitted use, you will need to obtain permission directly from the copyright holder.

Chapter 8
Overall Comparison of Energy Carriers

8.1 Technical Assessment of Liquefied Energy Carriers

The adoption of sustainable energy systems requires the implementation of effective methods for storing and transmitting energy. In this context, liquefied energy carriers such as LNG, liquefied hydrogen, ammonia, methanol, DME, and formic acid are of prime importance. Conducting a comprehensive technical evaluation of these carriers necessitates a profound comprehension of the energy demands linked to their manufacturing, storage, and transportation. This section specifically examines the energy inputs, efficiencies, and losses associated with these processes, offering comprehensive insights substantiated by current research.

8.1.1 Energy Requirements for Production

The production of energy carriers is substantially shaped by the energy demands intrinsic to their production processes. This section examines the numerous energy inputs required for the manufacture of different energy carriers, such as LNG, liquid hydrogen, ammonia, methanol, DME, and formic acid. Each carrier is manufactured by methods that are significantly different regarding their thermodynamic problems and technological complexities, which therefore determine their total energy efficiency and cost viability.

The LNG process necessitates cooling natural gas to an exceedingly low temperature of − 162 °C. The cryogenic liquefaction process is energy-intensive due to its several phases of gas compression and cooling. Advancements in liquefaction technology, including the adoption of mixed refrigerant and cascade cycles, have enhanced efficiency, consequently decreasing the relative energy consumption necessary for the process.

Conversely, the production of liquid hydrogen requires significantly more energy. Hydrogen liquefaction necessitates cooling to significantly lower temperatures, leading to energy usage that may constitute a considerable fraction of the lower heating value (LHV) of hydrogen. Moreover, the generation of hydrogen through water electrolysis is energy-intensive, and the ensuing liquefaction process further escalates the total energy consumption. The dual-stage energy demand renders liquid hydrogen one of the most difficult energy carriers to produce efficiently.

The production of ammonia by the Haber–Bosch process poses distinct energy challenges. The process is significantly affected by the energy demands linked to hydrogen feedstock production, primarily via natural gas reforming. The substantial energy need of this feedstock manufacturing process, which constitutes the majority of total energy consumption, considerably affects the efficiency of ammonia production. While alternate approaches such utilizing green hydrogen from electrolysis present possible environmental advantages, they also necessitate increased energy inputs.

The production of methanol, which generally entails the creation of synthesis gas by steam methane reforming and subsequent catalytic conversion, necessitates considerable energy input. The process can be enhanced by integrating innovative strategies, such as utilizing collected CO_2 and green hydrogen, although these methods generally elevate overall energy usage. Nonetheless, methanol synthesis continues to be competitive owing to its very advantageous energy efficiency relative to other chemical processes.

The synthesis of DME, whether via indirect methods such as methanol dehydration or direct synthesis from synthesis gas, exemplifies the intricate equilibrium between supplementary processing stages and energy efficiency. The dehydration phase in indirect synthesis increases energy requirements, whereas direct synthesis presents an alternate pathway that may enhance energy efficiency; however, it necessitates a higher capital expenditure for specialized apparatus.

The production of formic acid, usually achieved via the hydrolysis of methyl formate, is intrinsically energy-intensive. Innovative techniques, like CO_2 hydrogenation utilizing green hydrogen, are being investigated to establish carbon–neutral manufacturing pathways. Nonetheless, these novel solutions presently require increased energy inputs owing to the demands of electrolysis, highlighting a larger difficulty in aligning environmental sustainability with energy efficiency.

These production paths highlight the essential importance of energy input in the efficiency and viability of manufacturing various energy carriers. This section seeks to elucidate the energy requirements and technological factors of each process, so offering a thorough comprehension of the trade-offs and problems inherent in maximizing energy carrier production.

8.1.1.1 LNG

The liquefaction of natural gas is a highly energy-intensive procedure mostly because it requires the gas to be cooled to $-162\ °C$. In general, the energy needed for the

8.1 Technical Assessment of Liquefied Energy Carriers

process of liquefaction is approximately 10–15% of the energy present in the gas itself [1]. This method entails the compression of the gas and subsequent cooling in several stages using refrigerants, with the energy consumption being contingent upon the effectiveness of the liquefaction technology employed. Contemporary facilities utilize advanced technologies such as the mixed refrigerant cycle and the cascade cycle to enhance efficiency, resulting in energy consumption that is as low as 7–8% of the energy content of the gas [2].

8.1.1.2 Liquefied Hydrogen

Manufacturing liquefied hydrogen is a very energy-intensive process, as the liquefaction procedure consumes around 30–40% of the LHV of the hydrogen [3]. Production of hydrogen by water electrolysis necessitates around 50–55 kWh/kg of hydrogen, which is equivalent to around 150–165 MJ/kg. Liquefaction enhances the hydrogen's energy content by 10–13 kWh/kg [4], which is equivalent to 30–40% of its LHV. Therefore, the total energy expended in the production of liquefied hydrogen can amount to 65–70 kWh/kg, rendering it less efficient in comparison to alternative carriers.

8.1.1.3 Ammonia

The Haber–Bosch process for ammonia synthesis requires around 28–35 GJ per ton of final ammonia produced. The hydrogen feedstock is commonly obtained from the process of natural gas reforming, which constitutes around 70–85% of the overall energy consumption [5]. Although the use of green hydrogen from electrolysis may lead to higher energy demands, it effectively decreases carbon emissions. Ammonia plants of today attain energy efficiencies ranging from 60 to 65% depending on the LHV of the ammonia they produce [6].

8.1.1.4 Methanol

Methanol production from natural gas entails the process of SMR to generate synthesis gas, which is then subjected to catalytic conversion to methanol. The process requires around 25–35 GJ per ton of methanol [7]. Novel techniques employ captured CO_2 and green hydrogen, which can augment energy inputs through electrolysis but improve sustainability. The overall energy efficiency falls within the range of 65–70% [8].

8.1.1.5 DME

DME is synthesized by two main approaches: indirect synthesis including methanol dehydration and direct synthesis using synthesis gas. Indirect synthesis requires energy consumption comparable to that of methanol production, with an extra dehydration step totaling around 2–3 GJ per ton of DME. Direct synthesis is somewhat more energy-efficient, consuming around 28–38 GJ per ton. Typical process efficiencies vary between 60 and 68% [9].

8.1.1.6 Formic Acid

Conventional formic acid synthesis by methyl formate hydrolysis is a highly energy-intensive process, typically needing around 10–12 GJ per ton. Cutting-edge techniques include CO_2 hydrogenation with green hydrogen, which results in higher energy requirements for electrolysis but provides carbon–neutral routes. Present processes exhibit efficiencies ranging from 50 to 60%, but are anticipated to enhance with technological progress [10].

8.1.2 Energy Requirements for Storage

The storage of energy carriers has many issues inherently associated with the physical and chemical features of each substance. This section analyzes the energy demands related to the storage of different energy carriers, such as LNG, liquefied hydrogen, ammonia, methanol, DME, and formic acid. Every carrier requires a customized storage strategy that markedly impacts its total energy usage and operational expenses.

For cryogenic materials like LNG and liquid hydrogen, it is essential to sustain extremely low temperatures. LNG must be maintained at − 162 °C, requiring continuous energy input to counteract heat intrusion and regulate the naturally occurring boil-off gas (BOG) throughout storage. Liquefied hydrogen, maintained at a temperature of − 253 °C, necessitates advanced insulation and active cooling systems, exacerbated by elevated boil-off gas rates and material issues such as hydrogen embrittlement. These variables collectively increase energy needs and complicate storage systems.

Conversely, transporters such as ammonia provide greater flexibility in storage conditions. Ammonia can be stored either under refrigerated conditions at moderate subambient temperatures or at room temperature under elevated pressure. Each approach presents a distinct equilibrium between energy consumption and capital expenditure—refrigerated storage necessitates constant cooling, whereas pressurized storage mandates durable containment devices. Likewise, methanol, DME, and formic acid are advantageous when stored under ambient conditions, which markedly decrease their direct energy demands. Nonetheless, these carriers still require energy

8.1 Technical Assessment of Liquefied Energy Carriers

input for auxiliary systems like vapor recovery or corrosion control, which are essential for maintaining safe and efficient long-term storage.

This section offers a thorough analysis of how the distinct storage requirements of each energy carrier—spanning from cryogenic cooling to minimal ambient storage—directly influence the energy consumption and technological intricacy of their respective storage systems.

8.1.2.1 LNG

Long-term storage of LNG necessitates constant energy input to sustain temperatures at $-162°C$. The production of boil-off gas (BOG) is unavoidable, representing approximately 0.05–0.1% of the stored volume on a daily basis [11]. BOG can be reliquefied or utilized for electricity production, so partially compensating for the expenses associated with storage energy.

8.1.2.2 Liquefied Hydrogen

The maintenance of hydrogen at $-253°C$ requires sophisticated insulation and uninterrupted energy supply. The rates for boiling-off gas are rather higher than those of LNG, ranging from 0.2 to 0.4% per day or even higher [12]. Certain systems utilize active cooling, resulting in an increase in energy consumption. Furthermore, hydrogen embrittlement necessitates the use of specific materials, so indirectly raising energy expenses.

8.1.2.3 Ammonia

Ammonia can be stored at room temperature with a pressure of 10 bar or cooled to $-33\ °C$ at atmospheric pressure. Cooling in refrigerated storage requires around 0.2–0.3 GJ per ton [12]. Although pressurized tanks decrease the need for refrigeration, they necessitate strong containment system.

8.1.2.4 Methanol

Methanol is stored in a liquid state at room temperature and pressure, so reducing the need for significant energy inputs. While safety devices such as vapor recovery units consume energy, the total energy requirements for storage are negligible [12].

8.1.2.5 DME

The storage conditions for DME are 5–7 bar pressure at room temperature. Although pressurization necessitates energy, the quantities involved are rather modest, ranging from 0.1 to 0.2 GJ per ton [12]. This storage infrastructure is analogous to that of LPG, capitalizing on pre-existing technologies.

8.1.2.6 Formic Acid

Stored as a liquid under ambient conditions of temperature and pressure, formic acid exhibits minimal direct energy consumption during storage. Nevertheless, energy inputs are crucial for the maintenance of materials and systems in order to avoid corrosion.

8.1.3 Energy Requirements for Transportation

Transportation is an essential component of the energy supply chain, directly impacting the overall efficiency and cost feasibility of energy carriers. The energy demands for transportation fluctuate considerably based on the physical characteristics of the carrier and the mode of transit—be it marine vessel, pipeline, or traditional tanker. Every energy carrier, including LNG, liquefied hydrogen, ammonia, methanol, DME, and formic acid, presents distinct problems and energy losses during transportation that must be meticulously addressed to enhance operational efficiency.

LNG transportation generally depends on specialized maritime carriers that use a fraction of the cargo as fuel in the form of boil-off gas. Moreover, the transportation of LNG through pipelines necessitates the installation of compression stations to mitigate friction losses and sustain flow, thereby requiring additional energy consumption. Liquefied hydrogen, maintained at cryogenic temperatures, undergoes increased boil-off rates and requires additional energy for loading and unloading, resulting in significant cumulative losses over extended distances. Conversely, ammonia, which may be transported at moderate pressure or at ambient conditions, exhibits comparatively smaller energy losses, while methanol and formic acid advantageously utilize traditional liquid transport methods that have little energy penalties. DME necessitates moderate pressurization similar to LPG, positioning it in an intermediate category, with its energy consumption affected by the unique requirements for sustaining container pressure throughout transfer.

This section examines the diverse energy inputs necessary for transportation, emphasizing the variations in losses—stemming from boil-off, compression, or pumping—across different carriers and transportation methods. Comprehending these distinctions is crucial for enhancing logistical efficiency and minimizing the overall energy footprint linked to the transportation of energy carriers.

8.1 Technical Assessment of Liquefied Energy Carriers

8.1.3.1 LNG

During maritime transportation, LNG carriers use a portion of their cargo as fuel, referred to as BOG. Around 2–3% of LNG cargo is utilized during extended transportation journeys. Pipeline transportation necessitates compression stations, which consume approximately 0.3–0.5% of the energy being transported per 100 km [12].

8.1.3.2 Liquefied Hydrogen

The boil-off gas rates during transportation can occasionally reach levels as high as 0.2–0.3% per day. Further energy is necessary for the processes of loading and unloading. Typically, the overall energy losses in transportation can range from 10 to 13% when covering substantial distances [12].

8.1.3.3 Ammonia

The energy consumption of ammonia carriers is comparable to that of LPG carriers, when losses amount to less than 0.1% per day [12]. Low-pressure operations in pipeline transport result in an energy consumption of approximately 0.2–0.4% per 100 km compared to natural gas.

8.1.3.4 Methanol

The transportation of methanol is facilitated by conventional tankers, which incur minimal energy losses. Transport losses are minimal, amounting to less than 0.1%, mostly linked to pumping and handling phases [12].

8.1.3.5 DME

The transportation of DME necessitates the application of moderate pressure in containers that are analogous to those used for LPG. The pressurization and transfer processes use approximately 0.1–0.2 GJ per ton respectively.

8.1.3.6 Formic Acid

Standard liquid transport methods are employed for the transportation of formic acid using conventional chemical tankers. The energy consumption is negligible, mainly linked to pumping and handling activities.

8.1.4 Comparison and Analysis

The energy demands for generation, storage, and transportation differ markedly among energy carriers, highlighting the intrinsic thermodynamic, material, and procedural problems linked to each substance. An exhaustive analysis indicates that the total energy footprint of an energy carrier is influenced by both the efficiency of its production process and the subsequent energy requirements for stability during storage and effective transportation delivery.

LNG and liquefied hydrogen exemplify two extremes of energy intensity in manufacturing. The manufacturing of LNG, necessitating the cooling of natural gas to − 162 °C, is intrinsically energy-intensive owing to the multistage compression and refrigeration processes needed. Advancements in liquefaction technologies, including mixed refrigerant and cascade cycles, have allowed contemporary plants to decrease energy usage compared to the theoretical minimum. Conversely, the manufacture of liquid hydrogen has significantly greater obstacles. The procedure requires cooling to − 253 °C and entails significant energy expenditures for both liquefaction and hydrogen production, particularly when utilizing water electrolysis. This leads to an energy expenditure that can constitute a substantial fraction of hydrogen's lower heating value, rendering it one of the most energy-intensive processes among the carriers examined.

The cryogenic characteristics of LNG and liquid hydrogen highlight the significant energy demands associated with storage. LNG storage necessitates a constant energy supply to sustain very low temperatures and to regulate boil-off gas, although at comparatively lower rates than hydrogen. Liquefied hydrogen, due to its lower storage temperature, experiences elevated boil-off rates and requires enhanced insulation and active cooling systems. In contrast, carriers such as ammonia, methanol, and formic acid are advantageous due to their more lenient storage requirements. Ammonia can be stored in refrigerated tanks at low subambient temperatures or in pressurized containers at ambient temperature, each technique presenting unique trade-offs regarding energy input and infrastructural demands. Methanol and formic acid, generally maintained under ambient conditions, entail negligible direct energy expenses during storage; however, they necessitate energy for supplementary systems like vapor recovery and corrosion management. DME, maintained at modest pressures similar to liquefied petroleum gases, holds an intermediary status, where energy costs are mostly linked to sustaining pressured conditions rather than cryogenic temperatures.

Transportation further highlights the disparities among these energy transporters. LNG is generally conveyed by specialized marine vessels that utilize a segment of the cargo as fuel via the regulated usage of boil-off gas. The transportation of LNG by pipeline necessitates energy-intensive compression stations, while these expenses are distributed across the distance traveled. Liquefied hydrogen, because of its low storage temperature and elevated boil-off rates, incurs significant cumulative energy losses throughout transportation. The energy necessary for loading and unloading exacerbates these losses, rendering the transportation of liquefied hydrogen very

8.1 Technical Assessment of Liquefied Energy Carriers   147

difficult. Ammonia, methanol, and formic acid possess established transportation techniques that entail comparatively modest energy costs. Ammonia is delivered using systems similar to those for LPG, but methanol and formic acid employ standard tankers with negligible pumping and handling losses. DME necessitates pressurized containers and exhibits moderate energy requirements during transfer but is less energy-intensive than cryogenic carriers.

In summary, the comparative examination of energy demands across the manufacturing, storage, and transportation phases indicates that energy carriers like LNG and liquefied hydrogen necessitate substantial energy inputs at each stage of the value chain owing to their rigorous thermodynamic requirements. This affects both their total energy efficiency and the economic and environmental feasibility of these carriers in global energy markets. Conversely, carriers such as ammonia, methanol, DME, and formic acid, which require less stringent production and storage conditions, exhibit reduced energy consumption profiles. The disparities underscore the necessity of optimizing every phase of the energy carrier supply chain—via technology advancement and process integration—to improve overall efficiency and diminish the environmental impact of energy transportation. Comprehending these trade-offs is crucial for policymakers, engineers, and industry stakeholders aiming to create sustainable and economically feasible energy solutions in a progressively energy-aware society.

8.1.5 Case Study 1: Technical Assessment of Energy Carriers' Entire Supply Chain

This case study is adapted from [11]. This study analyzes the supply chain of three energy carriers: LNG, ammonia, and methanol, all of which originate from natural gas. The supply chain comprises multiple critical stages, beginning with the extraction and production of natural gas, followed by its conversion into one of three energy carriers, storage at the departure port, maritime transit, and ultimately, storage at the arrival port. Figure 8.1 depicts the comprehensive supply chain process from production to the target port.

8.1.5.1 Description

Natural gas is initially conveyed from extraction sites to specialized processing facilities, each tailored for a distinct energy carrier. At LNG facilities, natural gas is converted into a liquid state to facilitate efficient transportation. Ammonia plants transform natural gas into ammonia, which is subsequently cooled to its boiling point for liquid storage. Likewise, methanol plants transform natural gas into methanol, which retains its liquid state under standard conditions. Following manufacture, the energy carriers are held in cryogenic tanks at the departure port for roughly

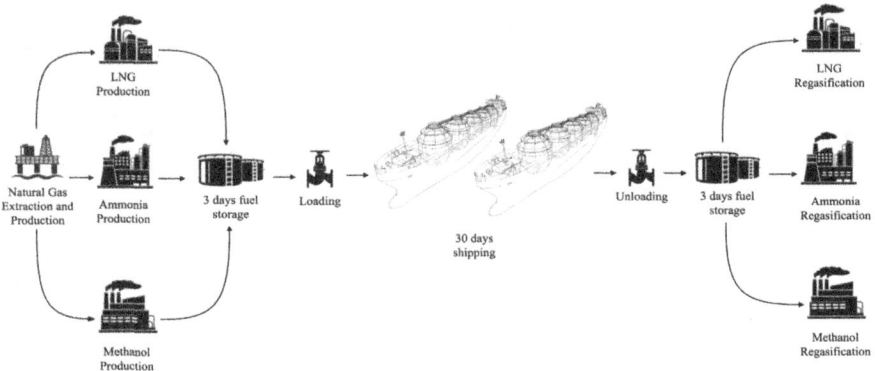

Fig. 8.1 Supply chain of the selected energy carriers including storage and ship transport [11]

three days. This storage duration encompasses the time necessary for port clearance and loading preparations. The storage tanks, with volumes ranging from 80,000 to 160,000 m^3, include stainless steel interiors for low-temperature resilience and concrete-reinforced exteriors to prevent leaks and maintain structural integrity.

Upon the arrival of the transport vessel, the liquid fuels are conveyed from the storage tanks to the ship utilizing high-capacity loading pumps. The loading rate fluctuates between 12,000 and 14,000 m^3 per hour, contingent upon the dimensions of the tanks. Each tank is filled to 98% of its total capacity to mitigate the possibility of spillage. The vessels utilized for transport are fitted with insulated tanks in their hulls to preserve the liquid fuels at necessary low temperatures. The predominant storage capacities of these vessels range from 120,000 to 160,000 m^3, with certain larger vessels attaining capacities of up to 270,000 m^3. These substantial tankers are predominantly utilized for long-range transport. The estimated shipping period in this study is 30 days, derived from the average transit time for LNG shipments over 160,000 m^3, exemplified by routes between Qatar and Japan.

Upon reaching the destination port, the liquid fuels are discharged into storage tanks, where they are retained for roughly three days prior to subsequent processing or distribution. Both LNG and ammonia are subsequently transported to a regasification facility, where they are transformed back into their gaseous states for industrial or energy applications. Conversely, methanol remains liquid due to its elevated phase transition temperature, which obviates the need for regasification. This organized supply chain guarantees the effective and regulated transit of LNG, ammonia, and methanol, enhancing their function as sustainable energy carriers.

8.1.5.2 Methodology

This case presents a technical evaluation of three energy carriers—LNG, ammonia, and methanol—through energy and exergy evaluations. This analysis assesses the energy demands at each phase of the supply chain and quantifies energy and exergy

8.1 Technical Assessment of Liquefied Energy Carriers

losses, especially those arising from boil-off gas (BOG) at various stages. The study analyzes the energy necessary for the conversion of natural gas into LNG, ammonia, and methanol, in addition to the energy required for loading, unloading, ship fueling, marine transportation, and, specifically for LNG and ammonia, regasification. The total energy required in each supply chain is ascertained by aggregating the energy demands of each step. Furthermore, BOG losses transpire in storage tanks, during loading and unloading, and throughout transit. The entire mass loss from BOG is assessed to facilitate a thorough comparison of the efficiency of LNG, ammonia, and methanol as energy carriers.

In this investigation, main ship parameters are constant. The vessel possesses a total volumetric capacity of 160,000 m^3, including four spherical tanks, each with a radius of 21.23 m. The environmental parameters are established at 1 bar of pressure and a temperature of 25 °C. Table 8.1 delineates the thermophysical properties of LNG, ammonia, and methanol at atmospheric pressure, derived from the Engineering Equation Solver (EES) program, which calculates these properties at designated temperatures and pressures.

By setting preset ship specifications and ascertaining the density of each liquid, the mass of each energy carrier is calculated at both the commencement and conclusion of the supply chain. This facilitates a comparative investigation of mass losses attributable to BOG for LNG, ammonia, and methanol. BOG is measured at each subprocess of the supply chain, guaranteeing a precise evaluation of losses at various stages. The energy can be calculated using the lower heating value (LHV) of each liquid, given the known mass of each energy carrier. Likewise, the accessible exergy is determined by the unique chemical exergy of each energy carrier. This method offers a systematic evaluation of the energy and exergy efficiency of LNG, ammonia, and methanol as transportable energy carriers.

The energy necessary to transform natural gas into LNG, ammonia, and methanol includes both direct energy inputs, such fuel consumption, and indirect energy inputs, such as power usage, per kilogram of the generated energy carrier. Every conversion process possesses unique energy requirements determined by its chemical and thermodynamic characteristics.

Table 8.1 Significant thermophysical properties of LNG, ammonia, and methanol

Thermophysical property	LNG	Ammonia	Methanol
Storage temperature (°C)	− 162	− 34	5
Density (kg/m^3)	423.1	682.8	805
LHV (MJ/kg)	48.6	18.6	19.9
Chemical exergy (MJ/kg)	51.7 [13]	22.4 [14]	19.84 [14]
Mass for single tank (kg)	16,924,000	27,312,000	31,640,000
Δh_{vap} (kJ/kg)	511.8 [15]	1371	1199
$\Delta h_{ex_{vap}}$ (kJ/kg)	234	150	41.6
U (W/m^2 k)	0.114 [16]	0.32 [17]	0.02 [18]

The liquefaction process for LNG utilizes refrigeration technology to cool natural gas to − 162 °C, converting it into a liquid condition. This method necessitates around 6.732 MJ/kg of energy [19]. The liquefaction process is crucial for diminishing the volume of natural gas, hence enhancing its suitability for storage and transit.

The synthesis of ammonia from natural gas transpires in two primary phases: hydrogen generation and ammonia synthesis. The initial phase entails the transformation of natural gas into synthesis gas, comprising carbon monoxide (CO) and hydrogen (H_2), by steam reforming. The shift reaction is employed to eliminate carbon monoxide and produce a combination of nitrogen (N_2) and hydrogen (H_2). In the second stage, ammonia is produced by the Haber–Bosch process, wherein hydrogen and nitrogen react in a fixed-bed reactor under elevated pressure and temperature, as illustrated by the chemical equation:

$$N_2(g) + 3H_2(g) \rightleftharpoons 2NH_3(g)$$

Following synthesis, refrigeration technology is utilized to liquefy ammonia at − 34 °C for storage and transportation. The total energy necessary for ammonia production from natural gas is roughly 28.656 MJ/kg [20]. This energy-intensive procedure is primarily ascribed to the high-pressure synthesis of ammonia and the supplementary cooling necessary for liquefaction.

The synthesis of methanol from natural gas has two principal phases. The initial phase involves the transformation of natural gas into synthesis gas via catalytic reforming, yielding a gas mixture of carbon monoxide (CO), carbon dioxide (CO_2), water (H_2O), and hydrogen (H_2). The second stage entails the catalytic synthesis of methanol, wherein synthesis gas is transformed into methanol via chemical processes. The subsequent equations delineate these two phases:

$$2CH_4(g) + 3H_2O(g) \rightarrow CO(g) + CO_2(g) + 7H_2(g)$$

$$CO(g) + CO_2(g) + 7H_2(g) \rightarrow 2CH_3OH(l) + 2H_2(g) + H_2O(g)$$

Methanol is stored as a liquid at 5 °C post-production. The total energy necessary for methanol production from natural gas is roughly 22.356 MJ/kg [21].

Subsequent to the production stage, the next phase in the supply chain is terrestrial storage. This research assumes no supplementary energy requirements for land storage at the departure and receiving terminals, as the storage tanks are adequately insulated and sufficient to sustain the necessary temperatures. In certain instances, refrigeration devices are employed to reduce boil-off gas (BOG) production [22]; nevertheless, some research suggest that BOG is instead flared into the atmosphere [23, 24]. This study examines the situation where BOG is flared without implementing any more efforts to mitigate its production. During loading and unloading, liquids are moved between storage tanks and ship tankers via pumps, necessitating approximately 0.00196 MJ/kg of energy [25].

8.1 Technical Assessment of Liquefied Energy Carriers

The transportation step entails the shipment of liquids from the departure port to the destination port. The energy necessary for the transportation of 160,000 m^3 of liquid is roughly 7.36 MJ/kg [25]. Upon arrival at the destination, LNG and ammonia are subjected to regasification, necessitating 0.512 MJ/kg [26] and 1.37 MJ/kg [15], respectively, while methanol retains its liquid state. The aggregate energy consumption across the supply chain is calculated by totaling the energy necessary for production, loading, transportation, unloading, and regasification, as shown in the subsequent equation:

$$E_{\text{consumption}} = \sum (E_{\text{production}} + E_{\text{loading}} + E_{\text{transporting}} + E_{\text{unloading}} + E_{\text{regasification}})$$

Energy consumption is the sum of energy production, energy loading, energy transporting, energy unloading, and energy regasification. Specific assumptions were established to do the analysis: (i) Energy and exergy losses from the extraction of raw gas to the production of natural gas are excluded from consideration. (ii) BOG produced during industrial processes and pipelines is eliminated from this analysis.

All three energy carriers—LNG, ammonia, and methanol—are stored and transported in tanks. Although ammonia and methanol present flammability and toxicity hazards, these issues can be alleviated with appropriate safety protocols. Both chemicals had extensive histories of industrial application, and their safety measures have been markedly enhanced over time. Ammonia and methanol rank among the most extensively manufactured and transported chemicals globally.

All three energy carriers are stored and transferred as cryogenic liquids, maintained below their respective boiling points. As full insulation is unattainable, heat transmission from the surrounding environment to the tanks transpires, resulting in the boil-off gas (BOG) phenomenon. As the absorbed heat elevates the liquid's temperature, a fraction of the surface liquid evaporates, leading to BOG. These phenomena transpire at several levels along the supply chain owing to the cryogenic characteristics of the energy carriers.

BOG generation occurs in land storage tanks, during loading and unloading operations, and throughout sea transit. The BOG generated in storage tanks is often either flared into the environment or directed to a reliquefaction unit; however, this study excludes the latter option. The intermittent loading of liquid into ship tankers also results in increased BOG production. Moreover, during transit, persistent heat infiltration into the shipping tanks leads to the formation of BOG [27]. As the majority of shipping tankers do not possess reliquefaction systems, the boil-off gas (BOG) is either released into the atmosphere or utilized as fuel for the vessel. Subsequently, supplementary BOG is generated when the liquid is discharged from the vessel and kept at the receiving terminal prior to use. Consequently, BOG production transpires at many pivotal stages along the supply chain, affecting the overall energy and exergy losses associated with each energy carrier.

The loading and unloading phase denotes the interval when a ship tanker is attached to the jetty at the receiving terminal and connected to a land storage tank using insulated pipelines. In this phase, boil-off gas (BOG) is produced as energy carriers are moved into and out of the storage tanks via pipelines. The BOG creation at this stage arises from two primary factors: (i) Thermal energy infiltrates the transfer pipe, resulting in energy losses as heat permeates the insulated conduit and (ii) cooling the energy carriers' heel and the ship's tank to align with the supply temperature of the energy carriers. This analysis assumes that the composition and temperature of the energy carriers remain constant during the loading and unloading procedures. The principal parameters utilized in the analysis of these stages are enumerated in Table 8.2.

The principal origin of BOG formation during loading and unloading is the infiltration of thermal energy into the transfer pipe. The energy lost by heat leaks is quantitatively expressed by the following equation, whereas the associated exergy loss is delineated by equation respectively [28].

$$q_{energy} = \frac{2\pi kL(T° - T^c)}{2.3\log_{10}\left(\frac{D+2t_h}{D}\right)}$$

$$q_{exergy} = \frac{2\pi kL(T° - T^c)}{2.3\log_{10}\left(\frac{D+2t_h}{D}\right)} \times \left|\left(1 - \frac{T°}{T_{boundary}}\right)\right|$$

where k represents conductivity, D denotes the inner diameter of the pipe, L signifies the length of the pipe, t_h indicates insulation thickness, $T°$ refers to ambient temperature, and T^c represents the temperature of energy carriers in the storage tank. $T_{boundary}$ represents the mean temperature between the medium and its surroundings. In the second component, the heat dissipated by the heel and the tank, determined by their heat capacities, is utilized to estimate the total BOG production during loading and unloading. Assuming the tank and heel temperatures are same, the energetic BOG

Table 8.2 Loading and unloading parameters

Parameters	Value
Stainless steel pipe thermal conductivity (W/m °C)	0.02
Pipe inner diameter (m)	0.038
Length of pipe (m)	2
Insulation pipe thickness (m)	0.04
Loading and unloading time (h)	13
Tank heat capacity (kJ/°C)	2.52 × 10
LNG isobaric specific heat capacity (kJ/kg °C)	3.477
Ammonia isobaric specific heat capacity (kJ/kg °C)	2.132
Methanol isobaric specific heat capacity (kJ/kg °C)	2.19
Heel mass	5% of the total mass
LNG heel temperature (°C)	− 160
Ammonia heel temperature (°C)	− 32
Methanol heel temperature (°C)	7

8.1 Technical Assessment of Liquefied Energy Carriers

is delineated in the following equation, while the exergetic BOG is articulated in another equation respectively [28].

$$\text{BOG}_{\frac{\text{loading}}{\text{unloading}_{\text{energy}}}} = \frac{tq_{\text{energy}} + (\text{HC}_T + C_p M_H)(T^H - T^c)}{\Delta h_{\text{vap}}}$$

$$\text{BOG}_{\frac{\text{loading}}{\text{unloading}_{\text{exergy}}}} = \frac{tq_{\text{exergy}} + (\text{HC}_T + C_p M_H)(T^H - T^c)}{\Delta h_{\text{ex}_{\text{vap}}}}$$

where t is either loading time or unloading time, HC_T is the tank heat capacity (kJ/°C), C_p is of the energy carriers isobaric specific heat capacity (kJ/kg °C), M_H is the mass of the heel (kg) retained at starting of loading or unloading, T^H is temperature of heel (°C).

A land storage period is required between the loading and unloading of tanks. The liquid is contained in a cryogenic storage vessel. The primary origin of boil-off gas during this period is the thermal influx into storage from the surrounding environment. The heat influx from the surroundings indicates the continuing generation of boil-off gas in the tanks. Due to the temperature disparity between the medium and the environment, heat infiltration manifests via the floor, walls, and roof of storage tanks, as illustrated in Fig. 8.2. The heat intrusion correlates with the tank's surface area. The tank's surface area is expressed in the following equation:

$$A = 4\pi r^2$$

The total energetic heat transfer into the tank is found by following equation while the total exergetic heat transfer into the tank is calculated as in the following equation, respectively:

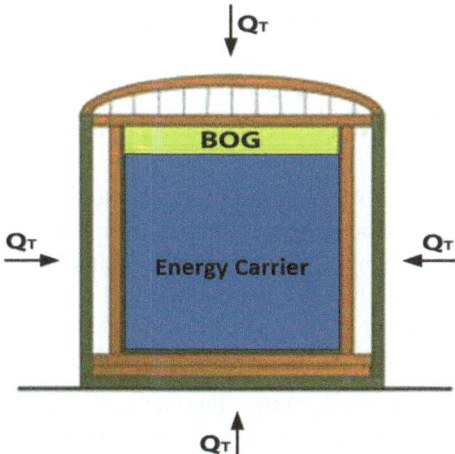

Fig. 8.2 Heat ingress into the energy carrier storage tanks

$$\dot{Q}_{\text{energy}} = UA\Delta T$$

$$\dot{Q}_{\text{exergy}} = UA\Delta T \times \left| \left(1 - \frac{T_\circ}{T_{\text{boundary}}} \right) \right|$$

where $\dot{Q}$ is the heat transfer in W, U is the overall heat transfer coefficient for the tank, A is the surface area in m^2 and ΔT is the temperature difference between the tank and the environment in K. The energy carrier evaporation is the ratio between the heat transfer into the tank and the heat of vaporization (Δh_{vap}) as represented in following equation for energetic value and the exergy carrier evaporation is the ratio between the heat transfer into the tank and the exergetic heat of vaporization ($\Delta h_{\text{ex}_{\text{vap}}}$) as in following equation, respectively:

$$\text{Energy carriers}_{\text{Evaporation}} = \frac{Q_{\text{energy}}}{\Delta h_{\text{vap}}}$$

$$\text{Exergy carriers}_{\text{Evaporation}} = \frac{Q_{\text{exergy}}}{\Delta h_{\text{ex}_{\text{vap}}}}$$

The energetic BOG and exergetic BOG can be determined by following equation and following equation, respectively:

$$\text{BOG}_{\text{land,energy}} = \frac{\text{Energy carriers}_{\text{evaporation}}}{\text{Energy carriers}_{\text{storage}}}$$

$$\text{BOG}_{\text{land,exergy}} = \frac{\text{Exergy carriers}_{\text{evaporation}}}{\text{Exergy carriers}_{\text{storage}}}$$

where boil-off gas is the gas boiled from the tank. Energy carriers storage is the available storage amount in the tank [16].

The voyage mode denotes the transportation phase during which energy carriers are transported from the departure terminal to the receiving terminal. During this interval, the generation of boil-off gas (BOG) may transpire due to various factors [23]:

- Heat transfer enters the tank due to temperature differentials between the cargo tanks and the surrounding environment.
- Cooling of the ship's tanks during ballast voyages, which entails prepping vacant tanks for the subsequent filling cycle.
- The sloshing of cargo, generated by wave motions, can result in the transfer of kinetic energy from the hull to the cargo tanks, so producing excess heat.

Among these causes, thermal infiltration is the principal cause of BOG formation during maritime transit. In turbulent sea conditions, sloshing effects may marginally facilitate BOG development due to the energy transfer from the vessel's movement to the stored liquid. Nonetheless, this effect is deemed insignificant in this study.

8.1 Technical Assessment of Liquefied Energy Carriers

Likewise, the cooling of ship tanks during ballast journeys is overlooked, as the analysis assumes the absence of active BOG reduction procedures. Consequently, the study exclusively examines heat ingress as the principal factor influencing BOG formation on vessels [23].

The thermal influx into the tanks is determined using the identical equation utilized for terrestrial storage. The overall heat transfer coefficient varies due to differences in tank materials and insulating characteristics. The heat transfer coefficients employed for various energy carriers are: (i) ammonia ship tanks: 0.32 W/m^2 K, (ii) methanol ship tanks: 0.02 W/m^2 K, (iii) LNG ship tanks: 0.2 W/m^2 K (modified owing to variations in terrestrial and maritime tank materials) [16, 18]. The study assesses energy and exergy losses at each stage by estimating the total BOG produced across the supply chain. The cumulative boil-off gas (BOG) along the entire supply chain is represented by the subsequent equation:

$$\text{BOG}_{\text{total}} = \sum (\text{BOG}_{\text{loading}} + 2 \times \text{BOG}_{\text{land}} + \text{BOG}_{\text{shipping}} + \text{BOG}_{\text{unloading}})$$

This equation considers BOG generation during loading, storage, transportation, and unloading, offering a thorough evaluation of energy carrier losses across the supply chain.

8.1.5.3 Results

Energy and exergy assessments are critical instruments for evaluating the performance of various energy carriers, including LNG, ammonia, and methanol. This analysis quantifies the energy and exergy losses within the supply chain, emphasizing losses attributed to boil-off gas (BOG) generation. Previous study predominantly examines LNG BOG losses throughout different supply chain phases, including land storage, loading and unloading, and tanker shipping; however, there is a paucity of data about BOG generation for ammonia and methanol. The main aim of this study is to evaluate and quantify BOG production for ammonia and methanol throughout various stages of their supply chains.

Table 8.3 delineates a comparison of daily BOG generation rates for LNG, ammonia, and methanol, derived from the findings of this study and corroborated by published figures. Hasan et al. [23] performed a dynamic simulation of LNG BOG throughout loading, unloading, and transportation, whereas Morgan [29] assessed BOG generation during the terrestrial storage of liquid ammonia. The BOG rates for LNG in this study exhibit minor discrepancies from the findings of Hasan et al. attributable to differences in heel percentage rates, thermal conductivity, and heel temperature. Moreover, no previous research has documented BOG rates for methanol at various points of the supply chain. This study offers novel insights and enhances the existing literature by reporting BOG rates for the methanol supply chain.

Table 8.3 Comparison of BOG rates of LNG, ammonia, and methanol with literature

LNG BOG (daily)	This study (LNG)	[23]	This study (ammonia)	[29]	This study (methanol)
Land storage (%)	0.0757338	–	0.01	0.01	0.00032
Loading (%)	0.16957	0.161	0.022	–	0.01667
Shipping (%)	0.12	0.137	0.024	–	0.0005
Unloading (%)	0.16957	0.151	0.022	–	0.01667
Land storage (%)	0.0757338	–	0.01	0.01	0.00032

The overall daily energy BOG rates for LNG, ammonia, and methanol are 0.610%, 0.098%, and 0.034%, respectively. Among all energy carriers, LNG demonstrates the greatest BOG losses, mostly attributable to the substantial temperature differential between the tank's interior and the external environment, together with the extended transportation length. In marine shipping, these conditions lead to LNG losses of roughly 3.6% of its total volume, rendering it more susceptible to evaporation losses than ammonia and methanol. This pattern is explicitly illustrated in Fig. 8.3.

Correspondingly, the total daily exergetic BOG rates for LNG, ammonia, and methanol are 0.491%, 0.068%, and 0.032%, respectively. Methanol demonstrates the minimal exergetic BOG generation, attributable to the reduced temperature differential between its liquid state and the surrounding environment. The decreased temperature differential diminishes overall heat infiltration, thereby reducing evaporation losses.

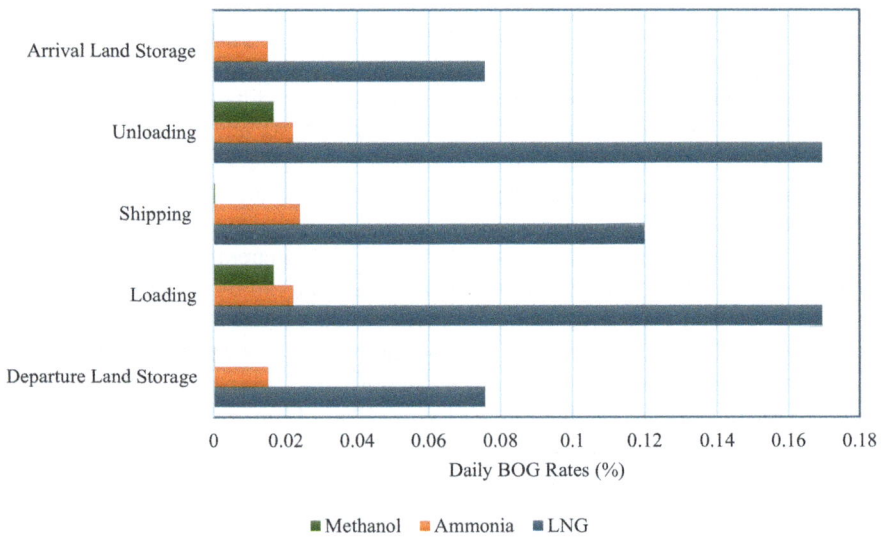

Fig. 8.3 Daily BOG rates of various subprocesses for LNG, ammonia, and methanol supply chain

8.1 Technical Assessment of Liquefied Energy Carriers

The daily BOG rates across several processes underscore that LNG constantly incurs the greatest mass losses across its supply chain, especially during shipping, loading, and unloading. Conversely, ammonia and methanol exhibit markedly reduced BOG formation rates, rendering them more stable during transport and storage.

The required energy in each process of the supply chain and the energetic and exergetic BOG values were analyzed and shown in Fig. 8.4.

The total energy requirements for LNG, ammonia, and methanol are 14.60 MJ/kg, 37.39 MJ/kg, and 29.71 MJ/kg, respectively. Among the three energy carriers, LNG has the lowest energy demand, primarily due to the efficiency of its liquefaction process. In contrast, ammonia requires the highest energy input, as its production involves air separation for nitrogen extraction and cooling below its boiling point (− 34 °C). Methanol, while simpler to handle after delivery, still has higher energy requirements than LNG, as its production process consumes nearly three times more energy than LNG liquefaction. This indicates that LNG is the most energy-efficient option for transporting natural gas; however, cost and environmental factors should also be considered in selecting the most suitable energy carrier.

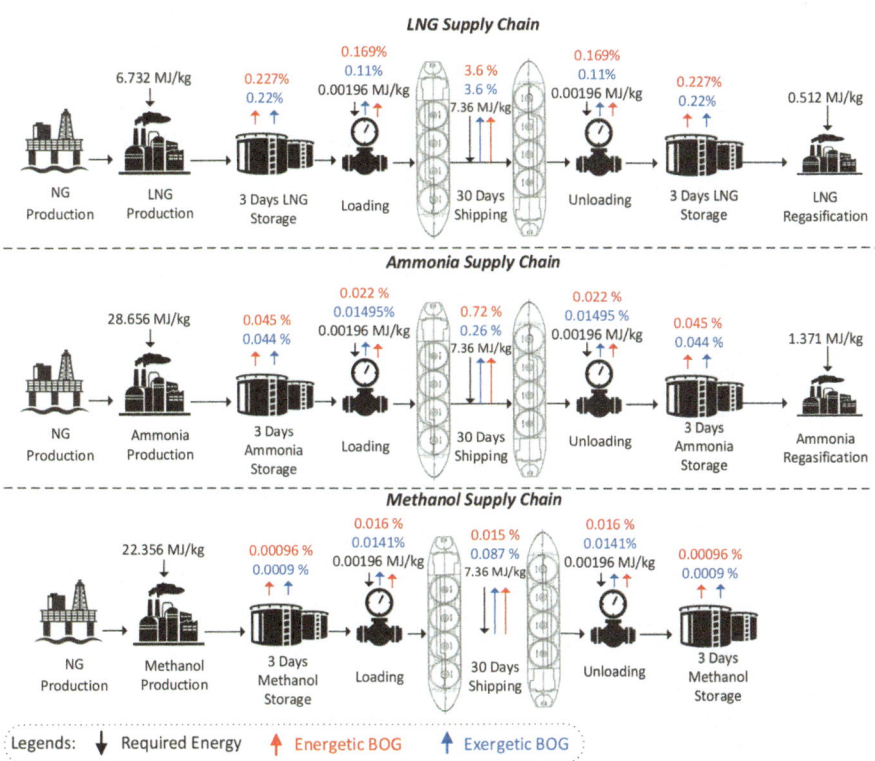

Fig. 8.4 Required energy and BOG rates for LNG, ammonia, and methanol [11]

With a fixed ship capacity of 160,000 m^3, the transportable masses of the three energy carriers are: (i) LNG: 67,696 tons, (ii) ammonia: 109,248 tons, and (iii) methanol: 128,800 tons. These values reflect the higher densities of ammonia (682.8 kg/m^3) and methanol (805 kg/m^3) compared to LNG, which has a lower density due to its cryogenic liquefaction process. These figures are presented in Table 8.4.

During the supply chain, boil-off gas (BOG) reduces the transported mass, impacting the overall efficiency of each carrier. For LNG and ammonia, over 80% of total BOG generation occurs during marine transportation, with other phases—such as loading, unloading, and storage—contributing less than 5%. In contrast, methanol exhibits a different BOG distribution, where 30% of its total BOG occurs during shipping, while 33% of losses happen during loading and unloading. This distinction highlights that transportation is the most critical phase for BOG generation in cryogenic carriers like LNG and ammonia, whereas methanol experiences the highest losses during handling due to its storage at near-ambient temperatures.

Energy carriers incur mass losses from boil-off gas (BOG) across their supply chains, including during land storage, loading, transportation, and unloading. Figure 8.5 illustrates the magnitude of these losses, presenting both energy-based and exergy-based losses for each supply chain.

Of the three energy carriers, LNG has the greatest mass loss, almost 5%, attributable to its cryogenic storage demands and the substantial temperature disparity between the liquid and its environment. Conversely, ammonia and methanol exhibit significantly reduced mass losses, remaining under 1%. The disparity is mainly due

Table 8.4 Total available and lost masses for LNG, ammonia, and methanol

	Energy carriers	LNG		Ammonia		Methanol	
	Analysis	Energy	Exergy	Energy	Exergy	Energy	Exergy
	Total mass (ton)	67,696	67,696	109,248	109,248	128,800	128,800
	Initial energy/exergy (TJ)	3.29	3.50	2.03	2.45	2.52	2.51
Total BOG	3 days' land storage (ton)	153.81	153.81	49.16	29.85	1.24	0.65
	Loading (ton)	114.79	74.47	24.03	10.12	21.47	9.55
	30 days' shipping (ton)	2437.06	2437.06	786.59	178.72	19.32	59.71
	Unloading (ton)	114.79	74.47	24.03	10.12	21.47	9.55
	3 days' land storage (ton)	153.81	153.81	49.16	29.85	1.24	0.65
	Delivered mass (ton)	64,721	64,802	108,315	108,989	128,735	128,719
	Delivered energy/exergy (TJ)	3.15	3.35	2.01	2.44	2.56	2.55
	Efficiency (%)	95.61	95.73	99.15	99.76	99.95	99.94

8.1 Technical Assessment of Liquefied Energy Carriers

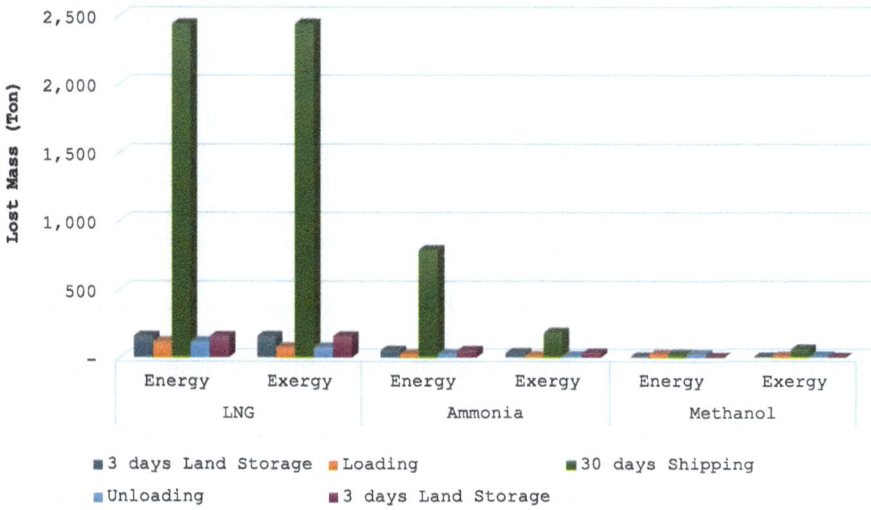

Fig. 8.5 Energetic and exergetic BOGs for LNG, ammonia, and methanol

to the elevated boiling temperatures of ammonia (− 34 °C) and methanol (5 °C), which diminish heat ingress and subsequent evaporation.

Exergetic losses are also maximum for LNG, roughly 3%, whereas ammonia and methanol exhibit minor exergetic losses. This results from the significant temperature disparity between LNG and ambient conditions, causing increased entropy formation and irreversible energy losses. The results highlight that LNG has the most substantial obstacles with cryogenic boil-off, whereas ammonia and methanol offer more stable storage and transportation conditions with reduced energy loss.

Future advancements, such as renewable methanol production and electrochemical ammonia synthesis, could reduce production energy demands, further enhancing their competitiveness. Additionally, ammonia and methanol could surpass LNG as preferred energy carriers if stricter environmental regulations or economic shifts favor zero-carbon or higher-value alternatives.

8.1.5.4 Conclusion

The conveyance of substantial energy volumes across extensive distances poses a significant problem in the global energy supply chain. Transporting energy in liquefied forms is one of the most effective alternatives, since it substantially decreases volume relative to gaseous forms, hence enhancing storage and transport efficiency. Nevertheless, liquefied energy carriers incur boil-off gas (BOG) losses attributable to temperature differentials between the stored liquid and the surrounding environment. This study provides a technical evaluation of three primary energy carriers produced from natural gas: LNG, liquid ammonia, and methanol. The assessment is performed using energy and exergy evaluations, offering insights into overall energy consumption, BOG-associated energy losses, and appropriateness for long-distance transport.

The total energy necessary for production and transportation is 14.60 MJ/kg for LNG, 37.39 MJ/kg for ammonia, and 29.71 MJ/kg for methanol. LNG exhibits the lowest energy consumption owing to the efficiency of its liquefaction process, but ammonia and methanol necessitate much more energy due to supplementary processing stages. Daily energy BOG losses are 0.610% for LNG, 0.098% for ammonia, and 0.034% for methanol.

Daily exergetic BOG losses are 0.491% for LNG, 0.068% for ammonia, and 0.032% for methanol. LNG experiences an around 5% loss of its total mass, while ammonia and methanol incur losses of less than 1%, rendering them more stable for long-distance transit.

Although LNG is the most energy-efficient choice regarding production energy requirements, it incurs the most losses from BOG, especially during transportation and storage. Notwithstanding this, LNG is extensively utilized because to its superior energy and exergy content per unit mass. Nonetheless, its enduring feasibility as an energy transporter is contingent upon economic and environmental considerations. Due to its carbon-free nature, ammonia has the capacity to supplant LNG in areas with stringent decarbonization regulations. Despite necessitating elevated energy inputs for manufacture, ammonia incurs negligible BOG losses, rendering it a more sustainable choice for long-distance transportation. Methanol exhibits the lowest BOG rates among the three carriers, indicating negligible mass losses during transmission. Moreover, methanol is more convenient to store and manage as it remains in liquid state at nearly ambient temperatures. Methanol may function as an efficient supplementary energy carrier in scenarios where stability and storage efficiency are paramount.

This study emphasizes the compromises among energy efficiency, BOG losses, and sustainability within the supply chain of LNG, ammonia, and methanol. LNG is the most energy-efficient carrier regarding production, although it incurs significant losses during transportation. Should market conditions change—such as decreasing LNG prices or more stringent environmental regulations—ammonia may become a feasible alternative owing to its carbon-free characteristics. Methanol, characterized by its minimal BOG losses, presents a viable alternative for areas seeking to diversify their energy transportation methods. The choice of the most appropriate energy carrier must take into account economic, technological, and environmental considerations, guaranteeing an ideal equilibrium among efficiency, sustainability, and long-term feasibility in global energy transport.

8.2 Economic Assessment

The economic feasibility of liquefied energy carriers is a crucial determinant affecting their acceptance in the global energy market. Comprehending the expenses related to the production, storage, and transportation of carriers like LNG, liquefied hydrogen, ammonia, methanol, DME, and formic acid is crucial for stakeholders involved in investment and policy formulation. This section offers a comprehensive economic evaluation of these carriers, emphasizing the capital and operational expenditures associated with their lifecycle, underpinned by current research.

8.2.1 Costs Associated with Production

The production of energy carriers entails intricate processes that necessitate considerable capital inputs and result in huge operational costs. This section analyzes the expenses related to the production of numerous primary energy carriers—specifically, LNG, liquefied hydrogen, ammonia, methanol, DME, and formic acid—using a qualitative approach. Each production pathway necessitates a distinct combination of advanced technology, specialized infrastructure, and energy-intensive processes, collectively influencing their economic profiles.

The production of LNG necessitates the construction of massive liquefaction facilities that function under cryogenic temperatures. These facilities are engineered to refrigerate natural gas to exceptionally low temperatures, a procedure that significantly increases capital expenditures and incurs substantial operational costs due to the energy demands of refrigeration. Conversely, the production of liquefied hydrogen presents distinct economic issues, as production systems must sustain even lower temperatures and manage increased energy requirements, resulting in heightened capital and operational costs.

The synthesis of ammonia, commonly accomplished using the Haber–Bosch process, represents an additional production method where cost determinants are intricately associated with the dual function of natural gas as both a feedstock and an energy source. The method requires substantial expenditures in reactor infrastructure and is highly susceptible to variations in feedstock prices, which subsequently affect total production costs. Methanol production, sourced from diverse feedstocks like natural gas or coal, demonstrates cost fluctuations based on the chosen production method and related environmental factors.

The production of DME introduces further complexity due to an additional processing step, usually involving the dehydration of methanol or direct synthesis from syngas, necessitating specialized equipment and heightened capital investment. The production of formic acid, whether via traditional hydrolysis processes or novel CO_2 hydrogenation procedures, is influenced by fluctuations in production costs determined by technology selections and energy consumption trends.

This introduction establishes the framework for a comprehensive examination of the economic forces that support the production of these varied energy carriers. By comprehending the interaction among capital investments, process energy demands, and feedstock expenses, both students and professionals can acquire a thorough understanding of how production economics affect the feasibility and competitiveness of each energy carrier within the wider energy framework.

8.2.1.1 LNG

The production of LNG necessitates substantial capital investment in liquefaction facilities, which may vary from $200 to $2000 per ton of annual capacity [30]. Operating expenses are affected by the energy required to cool natural gas to −162 °C, representing approximately 10–15% of the gas's energy content. The expense of feedstock (natural gas) constitutes a significant factor, with prices fluctuating internationally. In 2023, natural gas prices fluctuated between $2 and $6 per million British thermal units (MMBtu), impacting overall production expenses [31]. The economies of scale in extensive liquefaction facilities diminish the per-unit cost, rendering LNG competitive in global markets.

8.2.1.2 Liquefied Hydrogen

The production of liquefied hydrogen presents economic challenges due to substantial energy demands and capital expenditures. The capital expenditure (CAPEX) for hydrogen liquefaction facilities is projected to be between $50 million and $800 million for capacities of approximately 6000 kg/day to 200,000 kg/day, respectively. Operating expenses are influenced by electricity usage, with liquefaction requiring approximately 10–13 kWh per kilogram of hydrogen. At electricity rates of $0.05–$0.10 per kWh, the liquefaction expense ranges from $0.50 to $1.30 per kilogram. The total production cost of liquefied hydrogen, when factoring in hydrogen production costs through electrolysis ($4–$6 per kilogram), can amount to $5–$7 per kilogram, positioning it as one of the more costly energy carriers [32].

8.2.1.3 Ammonia

The synthesis of ammonia through the Haber–Bosch process necessitates significant capital expenditure, with the costs for constructing new facilities estimated between $1000 and $1500 per ton of annual capacity. Operating expenses are significantly impacted by the cost of natural gas, which functions as both a feedstock and an energy source. Natural gas constitutes roughly 70–85% of the production expenses. With a natural gas price of $4 per MMBtu, the production cost of ammonia varies between $300 and $400 per ton. The use of green hydrogen derived from electrolysis elevates production costs owing to increased electricity expenses, which may result in ammonia prices reaching $600–$800 per ton [33].

8.2.1.4 Methanol

Methanol production involves capital costs of $800–$1000 per ton of annual capacity for new plants. Operating expenses are contingent upon feedstock prices, chiefly natural gas or coal. At a natural gas price of $4 per MMBtu, the production costs for methanol vary between $250 and $350 per ton. The production of methanol from coal may be more cost-effective in areas with low coal prices, yet it poses environmental issues. The utilization of captured CO_2 and green hydrogen elevates production costs owing to increased energy expenditures, potentially attaining $400–$500 per ton [34].

8.2.1.5 DME

The production of DME through methanol dehydration incurs supplementary capital expenditures, estimated between $900 and $1200 per ton of annual capacity. Operating expenses encompass feedstock methanol costs and energy utilization for the dehydration procedure. With methanol priced at $300 per ton, the production costs of DME fluctuate between $400 and $500 per ton. The direct synthesis from synthesis gas can marginally decrease costs but necessitates a greater capital investment in specialized apparatus [35].

8.2.1.6 Formic Acid

The conventional method for producing formic acid through the hydrolysis of methyl formate incurs capital expenditures ranging from $1500 to $2000 per ton of annual capacity. Operating expenses are affected by the prices of methanol and carbon monoxide, with production costs varying between $500 and $700 per ton. Innovative CO_2 hydrogenation methods elevate capital expenditures due to sophisticated technological demands and increased operational costs stemming from electricity consumption for electrolysis, potentially escalating production costs to $800–$1000 per ton [36].

8.2.2 Costs Associated with Storage

The efficient storage of energy carriers is essential in contemporary energy systems, influencing both economic feasibility and operational efficacy of energy infrastructure. This section presents a thorough qualitative study of the expenses related to the storage of various energy carriers, including LNG, liquefied hydrogen, ammonia, methanol, DME, and formic acid. Each substance poses distinct storage issues primarily dictated by its physical and chemical characteristics, along with the conditions necessary to preserve it in a functional state.

Energy carriers like LNG and liquid hydrogen required cryogenic storage, requiring sophisticated containment systems to sustain extremely low temperatures. These conditions impose substantial technical requirements, leading to elevated capital investments and heightened operational costs, principally due to energy consumption for refrigeration and active cooling. Conversely, ammonia and methanol can be stored under milder conditions, either by moderate refrigeration or at ambient temperatures and pressures, often leading to reduced cost profiles. DME and formic acid occupy an intermediary position, each necessitating storage needs that reconcile technological complexity with cost-effectiveness.

This section aims to clarify the relationship between storage technology and economic cost determinants, emphasizing how variables like as temperature, pressure, material characteristics, and safety standards affect both capital and operational costs. Through the analysis of these characteristics, students and practitioners can attain a more lucid comprehension of the many problems and opportunities in the design and operation of storage systems for various energy carriers. The debate seeks to enhance understanding of the economic factors that must be considered in the development of sustainable and resilient energy storage technologies.

8.2.2.1 LNG

LNG storage needs specialized cryogenic tanks engineered to sustain exceptionally low temperatures. Cryogenic storage necessitates substantial financial investment in sophisticated containment systems designed to endure extreme operating conditions. Furthermore, the operational expenses are influenced by the necessity for constant refrigeration to regulate boil-off gas (BOG). Efficient management of BOG—via its application for power generation or reliquefaction—can alleviate certain operational costs; however, the overall system continues to be energy-intensive.

8.2.2.2 Liquefied Hydrogen

Storing liquefied hydrogen presents greater challenges than LNG storage because of the significantly lower temperatures necessary for liquefaction. The severe cryogenic conditions require storage methods that are both durable and technologically sophisticated, leading to significantly increased capital expenditures. Moreover, operational costs are typically high due to heightened boil-off rates and substantial energy requirements for active cooling systems. The requirement for specialist materials to mitigate hydrogen embrittlement additionally increases the total cost.

8.2 Economic Assessment

8.2.2.3 Ammonia

The storage of ammonia is typically more cost-effective than that of LNG and liquefied hydrogen. Storing at fairly low temperatures in refrigerated tanks results in reduced capital investments and reasonably low operating expenses. Ammonia may also be kept under pressured settings at room temperature. This approach decreases refrigeration needs, while it requires more durable tank construction to securely hold the material at elevated pressures. The storage of ammonia constitutes a more equitable economic proposal.

8.2.2.4 Methanol

Methanol provides considerable benefits regarding storage expenses because it may be securely stored at ambient temperature and pressure settings. The straightforward storage needs result in significantly reduced capital and operational costs. Methanol storage systems predominantly utilize conventional tank designs and regular maintenance, hence reducing the cost burden linked to its storage.

8.2.2.5 DME

The storage of DME requires moderate pressurization. This requirement necessitates a compromise between the demand for pressure-resistant storage options and cost-effectiveness. Despite the necessity for the storage system to sustain a regulated pressure environment, the total capital and operational costs are very low. Energy usage is mostly linked to pressure maintenance and regular safety monitoring.

8.2.2.6 Formic Acid

Formic acid is generally stored in standard chemical storage tanks at ambient conditions, resulting in a comparatively cheap cost structure. The capital investment for formic acid storage is moderate, while the recurring operational expenses are negligible. Essential factors in the storage of formic acid encompass the implementation of robust corrosion prevention measures and safety regulations, as well as the establishment of a secure and economical storage system.

8.2.3 *Costs Associated with Transportation*

Transportation charges for energy carriers are affected by the capital investment needed for specialized vessels or pipelines and the operational costs incurred during travel. This section analyzes the transportation issues associated with different energy carriers, emphasizing the qualitative distinctions in capital and operational necessities.

8.2.3.1 LNG

The long-distance transfer of LNG is mostly executed via marine shipping utilizing specialized LNG carriers. These vessels must be outfitted with sophisticated cryogenic containment technologies to securely preserve LNG at exceptionally low temperatures. The capital investment for these vessels is considerable, indicative of the advanced technology and rigorous safety standards mandated. Operational expenses for LNG transportation are influenced by variables including trip distance, fuel costs, and charter agreements. For shorter distances, pipeline transportation is a more cost-effective alternative owing to reduced per-unit transport costs.

8.2.3.2 Liquefied Hydrogen

Transporting liquefied hydrogen poses distinct issues owing to its exceedingly low liquefaction temperature and elevated boil-off rates. The specialist tankers carrying liquid hydrogen necessitate substantial investment, as they must utilize modern materials and insulation methods to preserve freezing conditions. Furthermore, operational expenses are increased by the energy requirements of active cooling systems necessary to mitigate boil-off losses during transportation. Although pipeline transportation of hydrogen is occasionally contemplated, material constraints limit its extensive use, and even in viable scenarios, it incurs greater expenses per unit distance compared to traditional carriers.

8.2.3.3 Ammonia

Ammonia advantages from a well-established transportation infrastructure in the chemical sector. The boats utilized for ammonia transportation are less capital-intensive than those for LNG or liquid hydrogen, as ammonia may be stored under milder conditions. Transporting ammonia over extensive distances is comparatively cost-effective, and pipelines present a viable alternative owing to their efficient cost framework at intermediate lengths. The selection between maritime and pipeline transport often hinges on the particular logistical and economic circumstances.

8.2.3.4 Methanol

The transportation of methanol is marked by comparatively cheap capital and operational expenses. Standard chemical tankers are typically utilized for methanol, as it can be securely stored and transported under ambient circumstances. The existing infrastructure for methanol transport and handling enhances its cost efficiency by minimizing the necessity for specialist equipment or elaborate safety protocols necessary for more volatile or cryogenic materials.

8.2.3.5 DME

The conveyance of DME necessitates pressurized containers, akin to those employed for liquefied petroleum gases. Despite incurring elevated construction expenditures compared to systems for chemicals maintained at ambient settings, these pressurized systems are engineered to securely accommodate the distinct physical features of DME. Operational expenses for DME transport are affected by the necessity to sustain a controlled pressured environment; nevertheless, the capacity to modify existing LPG infrastructure might mitigate the total outlay.

8.2.3.6 Formic Acid

Formic acid is generally delivered using standard chemical tankers, utilizing storage at ambient conditions. Nonetheless, its corrosive properties need the implementation of specialist materials and improved safety protocols in the construction of conveyance apparatus. Although these features may marginally elevate both capital and operational expenditures relative to less reactive compounds, the total cost remains comparable to other chemicals with analogous handling properties.

8.2.4 Comparison and Analysis

The transportation of energy carriers is a complex issue that necessitates a delicate equilibrium between capital investments and operational costs, as each energy carrier poses distinct obstacles due to its intrinsic physical and chemical characteristics. In the comparison of LNG, liquefied hydrogen, ammonia, methanol, DME, and formic acid, various elements are crucial for comprehending the economics of transportation. LNG and liquefied hydrogen require cryogenic storage tanks engineered to sustain extremely low temperatures over extended distances. This requirement significantly increases capital expenditures, as the design and construction of such boats necessitate advanced materials, specialized insulation, and stringent safety requirements. Furthermore, the operational expenses for these carriers are exacerbated by the necessity for constant energy input to regulate boil-off gas, posing a considerable difficulty during prolonged journeys. Conversely, carriers like ammonia and methanol, which may be stored under milder temperature conditions or even at ambient temperature, benefit from a more developed and cost-effective transportation infrastructure, leading to reduced capital and operational costs.

The disparity between LNG and liquid hydrogen transportation is notably pronounced. Both necessitate cryogenic storage; but, the far lower liquefaction temperature of hydrogen presents more technical challenges. The sophisticated engineering required to maintain liquefied hydrogen necessitates that specialized tankers entail not only elevated original investments but also higher operations expenses due to increased boil-off and more stringent cooling demands during transport. This comparison emphasizes the influence of the thermophysical parameters of an energy carrier on its transportation logistics, illustrating the overarching principle that lower liquefaction temperatures typically result in elevated overall costs. Conversely, ammonia transportation utilizes the established infrastructure of the chemical sector. The manageable handling and storage of ammonia—whether in refrigerated environments or as a pressurized liquid at room temperature—leads to reduced capital expenditures and more cost-effective operating frameworks. Ammonia is a potentially appealing energy carrier for transportation, especially for long-distance shipments where safety and cost-effectiveness are critical.

The economic dynamics alter when contrasting methanol, DME, and formic acid. Methanol, carried via traditional chemical tankers, is advantageous when stored under ambient conditions. This fundamentally streamlines the design and operational requirements of its transportation system, resulting in considerably lower capital expenditures and minimum operational costs. DME need pressurized vessels akin to those utilized in the LPG sector, positioning it in a transitional phase; it requires more resilient systems than methanol yet does not attain the stringent specifications linked to cryogenic carriers. The adaptation of existing LPG infrastructure may provide cost savings; however, the requirement for pressurization inherently entails greater expenses than carriers that do not necessitate such precautions. Formic acid, while comparable to methanol in being stored under ambient settings, presents further complications owing to its corrosive properties. The materials employed for transport must be chosen for their corrosion resistance, which somewhat increases both capital and operational costs. Nonetheless, the comprehensive transportation cost framework of formic acid is rather reasonable contrasted to the more technically intricate cryogenic infrastructure necessary for LNG and liquefied hydrogen.

In addition to the immediate economic factors of vessel design and operational energy, these transportation systems must be assessed within a wider framework. Elements such as fuel expenses, regulatory mandates, and infrastructure availability significantly influence the whole cost structure of delivering various energy carriers. Pipeline transportation presents a viable option for specific carriers over limited distances; nonetheless, its practicality is significantly influenced by the compatibility of materials with the conveyed substance and the geographical setting. LNG pipelines can be economically viable over moderate distances; nevertheless, the constraints of material compatibility frequently inhibit the extensive usage of liquefied hydrogen pipelines. Ammonia pipelines, on the other hand, leverage substantial expertise in the chemical industry, rendering them a feasible choice for integrated supply chains.

The transportation costs of these energy carriers reflect the intricate relationship among physical properties, technical demands, and infrastructural capacities. LNG and liquefied hydrogen encounter significant obstacles to cost reduction owing to the stringent conditions required for secure transport, whereas ammonia and methanol provide more economically advantageous profiles due to their less demanding storage needs and existing handling infrastructures. DME and formic acid occupy middle positions, characterized by modest technological obstacles and specialized material requirements, resulting in a balanced yet significant transit cost. This analysis underscores the need for a customized strategy in the design and operation of transportation systems, aligning the distinct characteristics of each energy carrier with the existing technological and economic frameworks, thus enhancing both safety and cost efficiency in a swiftly changing energy environment.

8.2.5 Case Study 2: Economic Assessment of Energy Carriers' Entire Supply Chain

This case study is adapted from [37]. This study examines the problem of Boil-Off Gas (BOG) by investigating the transportation of natural gas in liquefied states. BOG refers to the gas that vaporizes during the storage and transit of liquid energy carriers as a result of heat absorption. This study treats BOG not merely as a technical challenge but as a unit cost, incorporating the energy loss from BOG into the comprehensive economic evaluation. This method offers numerous benefits. Initially, it assists in determining the most efficient liquid energy carrier by analyzing BOG-related mass losses among various carriers. Given that each energy carrier exhibits a distinct rate of BOG generation, opting for a carrier with reduced BOG costs can result in substantial economic and energy efficiencies. Furthermore, BOG management can be executed through two principal methods: enhancing tank insulation to minimize heat absorption or employing reliquefication devices that capture and reliquefy the evaporated gas. By assessing the cost of BOG across several energy carriers, enterprises can select the most cost-effective approach to address this issue.

8.2.5.1 Description

This study examines the viability of employing alternate energy carriers for large-scale energy transportation, in addition to cost-saving techniques. Historically, LNG has been the primary medium for transporting natural gas; however, alternatives such as liquid ammonia, methanol, DME, and liquid hydrogen are increasingly being considered. The study computes the aggregate production and transportation expenses of various carriers, using BOG generation as a separate cost element for a comprehensive comparison. It categorizes costs into three principal types: capital cost (investment in production and transportation infrastructure), operational costs (recurring charges for transportation and storage), and BOG cost (the economic repercussions of gas evaporation losses).

The study also analyzes the impact of natural gas prices on the overall production costs for various liquefied carriers. It evaluates the transportation cost per unit of volume, mass, and energy content for each energy carrier, providing a comprehensive comparison of their efficiency in energy transfer. The study examines the impact of variations in ship capacity and transit distance on the total cost of transporting LNG. The study analyzes these elements to offer useful insights into the most cost-effective and efficient means for transporting natural gas, hence assisting decision-makers in choosing the optimal liquefied energy carriers for future energy trading and distribution.

8.2.5.2 Methodology

Natural gas from Qatar's North Field, one of the world's largest non-associated natural gas reserves, is delivered to various processing plants in Mesaieed Industrial City (MIC), Qatar. These plants convert natural gas into LNG, liquid ammonia, methanol, DME, and liquid hydrogen. With recoverable reserves of 900 trillion cubic feet (tcf), Qatar plays a key role in maximizing the economic value of its natural gas resources by overseeing the exploration, conversion, and transportation of hydrocarbons [38]. Qatar gas, a state-owned company, manages major phases of the LNG supply chain, including the conversion of natural gas into LNG and its transportation to energy-importing countries [39].

This study focuses on the entire supply chain of liquefied energy carriers, covering both conversion and transportation. After the raw natural gas is processed into its liquefied forms, these energy carriers are transported to regions with high energy demand. To accurately evaluate costs, this study analyzes both the production and transportation costs of each energy carrier. The total production cost of liquefied energy carriers includes three key components:

1. Investment Cost—The capital required to build and maintain the production plants.
2. Operational Cost—The expenses associated with running the plants.
3. Boil-Off Gas (BOG) Cost—The cost associated with gas losses during production.

During production, BOG is either released into the environment or sent to a reliquefication facility for reprocessing [40]. Since BOG generation is a significant issue in the liquefied energy supply chain, this study proposes treating BOG losses as a cost component within the total production cost. Similarly, the total transportation cost is analyzed in the same manner. When liquefied energy carriers are shipped across oceans, BOG is generated due to heat absorption. The evaporated gas can either be used as fuel to power the ship or reliquefied onboard and returned to storage tanks [41]. Because different energy carriers manage BOG differently, this study accounts for lost mass due to BOG as a cost when calculating transportation expenses. Therefore, the total transportation cost consists of:

8.2 Economic Assessment

1. Capital Cost of the Ship—The investment required to build and maintain transport vessels.
2. Operational Cost—The expenses involved in running the ship.
3. BOG Cost—The economic impact of evaporated gas losses during transport.

By incorporating BOG as a unit cost, this study provides a more accurate representation of the economic efficiency of different energy carriers. The cost breakdown for producing and transporting one gigajoule of energy from natural gas to high-demand regions is presented using a discounted cash flow analysis, with all calculations expressed in US dollars as presented in Fig. 8.6.

Production Cost

To ascertain the manufacturing cost of liquid energy carriers, all relevant expenditures from various production facilities are evaluated. This study utilizes actual production data from facilities situated in the same region, Mesaieed Industrial City (MIC), Qatar, to guarantee an equitable comparison. These facilities transform natural gas

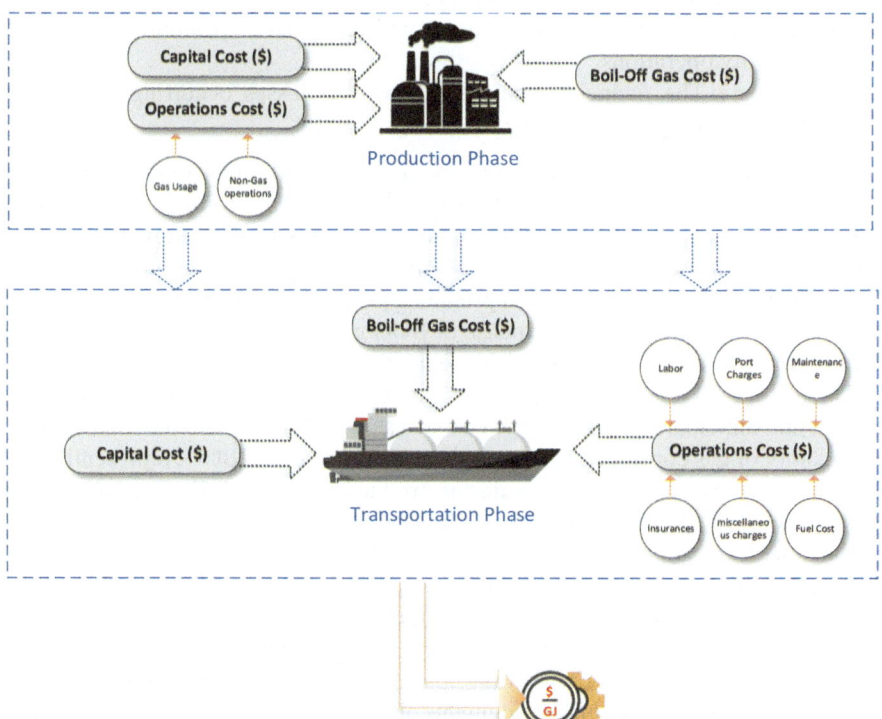

Fig. 8.6 Involved phases to determine the total cost of a liquefied energy carrier (from [37], licensed under CC-BY 4.0)

Table 8.5 Significant thermophysical properties of LNG, DME, ammonia, methanol, and hydrogen

Thermophysical properties	LNG	DME	Ammonia	Methanol	Hydrogen
Density (kg/m^3)	423.1	735.5	682.8	805	71.1
LHV (MJ/kg)	48.6	28.9	18.6	19.9	120
BOG during production (%)	0.176	0.024	0.052	0.071	1.189
Transportation BOG (%)	0.12	0.011	0.025	0.0005	0.52

Source Engineering Equation Solver [12, 42]

into liquefied energy carriers, such as LNG, liquid ammonia, methanol, DME, and liquid hydrogen. The production costs of each energy carrier differ due to their distinct physical and chemical features. Table 8.5 delineates essential thermophysical parameters, including density, lower heating value (LHV), and boil-off gas (BOG) generation rates for production and transportation.

The cost framework for generating liquid energy carriers has three primary components:

1. Capital cost refers to the investment required for constructing the plant, contingent upon its capacity and the period of construction.
2. Operational costs encompass fuel usage and additional expenditures associated with facility operations.
3. BOG Cost—The financial repercussions of gas losses occurring throughout production, storage, and loading.

In LNG production, approximately 50% of capital expenditures are allocated to gas liquefaction, while the remaining 50% is distributed among storage (18%), utilities (16%), loading facilities (10%), and gas pretreatment (6%) [43]. In the manufacture of liquefied ammonia, costs are significantly influenced by gas consumption, estimated at 25 MMBtu every ton of ammonia produced [44]. Due to the varying boil-off gas rates of each energy carrier, the expense of BOG generation is incorporated into the overall production cost in this analysis as shown in Table 8.6.

The investment cost is determined by the expenses associated with plant development and anticipated lifespan. The cost of gas usage is calculated by multiplying the quantity of gas consumed by its market price. The cost of BOG, representing the financial repercussions of gas losses during production, is determined by multiplying the volume of BOG produced by the selling price of the energy carrier. The entire production cost, calculated as the aggregate of all aforementioned expenses divided by the total energy generated, yields the cost per unit of energy in $/GJ. This study analyzes manufacturing costs with current literature to elucidate cost variations. A cost breakeven analysis is performed to assess the contributions of investment costs, operating costs, and BOG costs to the total production cost. This information is essential for making economical judgments about the reduction of BOG losses and the optimization of production techniques. The production cost of all five liquefied energy carriers, which utilize natural gas as feedstock, is directly influenced by variations in natural gas prices. A sensitivity analysis is conducted to comprehend this

8.2 Economic Assessment

Table 8.6 Production cost statistics for each liquefied energy carrier

	Unit	LNG	Ammonia	Methanol	DME	Hydrogen
Plant capacity						
Production[a]	kg/a	9,000,000,000	1,275,000,000	1,275,000,000	915,000,000	450,000,000
Production	kg/day	25,714,286	3,642,857	3,642,857	2,614,286	1,285,714
Capital cost						
Capital cost	$	5,225,500,000	605,710,000	378,600,000	390,710,000	378,200,000
Return on capital[b]	%	14.6	14.6	14.6	14.6	14.6
Investment cost	$	762,923,000	88,433,660	55,275,600	57,043,660	55,217,200
Operations' cost						
Non-gas cost	$	1,018,700,000	165,530,000	91,600,000	104,180,000	88,620,000
Gas usage	GJ/y	388,000,000	29,620,000	32,200,000	32,610,000	68,000,000
Cost of gas	$/GJ	2	2	2	2	2
Usage gas cost	$	776,000,000	59,240,000	64,400,000	65,220,000	136,000,000
BOG cost						
BOG from storage/loading	kg	15,813,000	663,000	310,080	649,650	5,350,500
Market price[c]	$/GJ	5.93 [45]	28.2 [46]	16.3 [47]	15.06 [48]	12 [49]
BOG cost	$	4,557,275	908,654.76	103,665.95	282,749.77	7,704,720
Total costs	$	2,562,180,275	314,112,315	211,379,266	226,726,410	287,541,920
Available energy	GJ	437,400,000	23,715,000	25,372,500	26,443,500	54,000,000

Source [44]
[a]This represents the production for 350 days in a year
[b]The discounted cash flow (DCF) rate of 10% for 3 years constructing duration and a plant lifetime of 20 years generates a return on capital investment of 14.6%
[c]Selling price for each commodity in the year of 2020

effect, assessing production costs across various natural gas price scenarios from 1 to 4 $/GJ. This analysis evaluates the impact of price fluctuations on the economic viability of each energy carrier, facilitating improved decision-making for future investments in liquefied energy production.

Table 8.7 Capital cost of LNG, liquid ammonia, methanol, DME, and liquid hydrogen ship in ($/m^3)

Energy carrier	Tanker cost ($/m^3)	References
LNG	1200	[50, 51]
Liquid ammonia	1016	[29]
Methanol	750	[50, 52]
DME	750	[50, 52]
Liquid hydrogen	1355	[29]

Transportation Cost

The overall expense of moving liquid energy carriers from supply areas to demand areas comprises three primary elements:

1. Capital Expenditure for Shipping Tankers—The financial outlay necessary to construct specialized tankers for each energy carrier.
2. Operational Costs—Expenditures associated with the operation and upkeep of vessels during transit.
3. Energy Losses (Boil-Off Gas—BOG Generation)—The quantity of gas that vaporizes during transit, signifying a reduction in transferred energy.

This study employs a cost-per-unit-volume ($/m^3) methodology to assess the capital cost of shipping tankers. Table 8.7 delineates the projected capital expenditures for various energy carriers, indicating the cost per cubic meter ($/m^3) of storage capacity for LNG, liquid ammonia, methanol, DME, and liquid hydrogen. The entire capital cost of a liquefied energy tanker can be determined by multiplying the cost per cubic meter by the total capacity of the tanker.

The expense of shipping tankers fluctuates according to the thermophysical characteristics of the energy carriers they convey. LNG must be kept at $-162\,°C$, necessitating high-quality insulating materials, hence elevating the overall cost of the tanker. This study assumes that all five shipping tankers—designed for LNG, liquid ammonia, methanol, DME, and liquid hydrogen—possess an identical capacity of 160,000 m^3 to facilitate a fair comparison. The energy capacity of each tanker is contingent upon the density of the energy carrier, whereas the transferred energy is quantified by its Lower Heating Value (LHV).

The overall transportation expense comprises three primary components: construction expenditure, operational expenses, and boil-off gas (BOG) expenditures as presented in Table 8.8. The capital cost of ships pertains to the investment necessary for constructing and maintaining a tanker for each energy carrier, determined by the cost per cubic meter of storage. The expense is projected for a 15-year vessel lifespan with a 10% discount rate. The operational expenses encompass crew pay, port fees, maintenance, insurance, and fuel use. The vessels utilize heavy fuel oil (HFO), and the aggregate fuel requirement is incorporated into operational expenses. A ship engine operating on HFO consumes 9.234 kg (9.419 L) of fuel to generate 100 kWh of power. A fully loaded tanker with an engine power of 31,400 kW consumes 2899 kg of HFO during its voyage [53, 54]. The mass loss varies due to the differing

8.2 Economic Assessment

evaporation rates of each energy carrier during transmission. The BOG cost is determined by multiplying the mass of lost energy by the market price of each energy carrier.

To facilitate a comprehensive comparison, the transportation cost of each energy carrier is computed in $/GJ (cost per unit of energy delivered) and $/kg (cost per unit mass transported). This facilitates a more accurate comparison of economic efficiency among various energy carriers. A sensitivity analysis is performed to elucidate variations in transportation costs by altering two critical parameters: ship capacity and transportation distance. The impact of augmenting tanker capacity from 100,000 to 160,000 m^3 and then to 250,000 m^3 is analyzed to assess the influence of economies of scale on overall transportation expenses. Vessels of greater size may diminish per-unit expenses owing to enhanced transportation efficiency. This study assesses transportation costs for liquefied energy carriers supplied from Qatar to significant energy-importing nations, considering various shipping distances: Japan (12,000 km), India (2400 km), and China (9700 km) [55]. Transportation costs fluctuate considerably based on the destination due to the increased fuel consumption and BOG losses associated with longer routes. This study analyzes these parameters to offer insights on optimizing shipping costs, facilitating the selection of the most economical energy carrier for long-distance transportation.

8.2.5.3 Results

The production cost of LNG, liquid ammonia, methanol, DME, and liquid hydrogen is calculated by summing the capital cost, operational cost, and the expenses related to boil-off gas (BOG) losses. The conversion of natural gas into these liquid energy carriers utilizes a minimum of 80% or more of the gas energy input. Alternative methods, including electrolysis, biomass conversion, and other sources, have been suggested for the creation of these energy carriers; however, their production efficiencies are often inferior to those achieved with natural gas. This renders natural gas the most economical primary fuel for extensive output [57, 58].

Table 8.9 compares the overall production costs of these liquid energy carriers as determined in this study and as documented in the literature. The production costs of LNG, liquid ammonia, methanol, DME, and liquid hydrogen exhibit minor discrepancies between the two sources, with LNG presenting the lowest cost among the carriers. Liquid hydrogen is comparatively cost-effective when assessed against liquid ammonia, methanol, and DME. The strong correlation between this study's findings and the available literature validates the accuracy of the cost estimations. Comparing several production methods reveals that natural gas is the most cost-effective technique for large-scale manufacturing of these liquid energy carriers.

The estimated production costs in this study are marginally elevated compared to those documented in the literature, as BOG expenses are explicitly included in the total production cost. Among the five energy carriers, LNG and liquid hydrogen have the lowest production costs, whereas liquid ammonia incurs the greatest cost

Table 8.8 Cost parameters for ocean transportation of LNG, ammonia, methanol, DME, and hydrogen from Qatar to Japan

	Unit	LNG	Ammonia	Methanol	DME	Hydrogen
Ship capacity	m^3	160,000	160,000	160,000	160,000	160,000
Capacity	kg	67,696,000	109,248,000	128,800,000	117,680,000	11,376,000
Logistics						
From Qatar to Japan						
Distance [55]	km	12,000	12,000	12,000	12,000	12,000
Sailing time[a]	day	13	13	13	13	13
Trips/year[b]		24.14	24.14	24.14	24.14	24.14
Sailing days/year	day	313.79	313.79	313.79	313.79	313.79
Capital costs						
Capital costs	$	192,000,000	162,560,000	120,000,000	120,000,000	216,784,000
ROC[c]	%	15.19	15.19	15.19	15.19	15.19
Total capital costs	%	29,164,800	24,692,864	18,228,000	18,228,000	32,929,490
Operations cost						
Labor	$	2,520,000	2,520,000	2,520,000	2,520,000	2,520,000
Required fuel	kg	7,327,540	11,825,205	13,941,549	12,737,900	1,231,359
Fuel cost[d,e]	$	4,249,973	6,858,619	8,086,099	7,387,982	714,188
Port charges	$	3,610,000	3,610,000	3,610,000	3,610,000	3,610,000
Maintenance (4% Capex)	$	7,680,000	6,502,400	4,800,000	4,800,000	8,671,360
Insurance (15% OpenX)	$	2,708,996	2,923,653	2,852,415	2,747,697	2,327,332
Misc (10% Opex)	$	1,805,997	1,949,102	1,901,610	1,831,798	1,551,555
Total operating cost	$	22,574,966	24,363,774	23,770,123	22,897,478	19,394,435
BOG cost						

(continued)

8.2 Economic Assessment

Table 8.8 (continued)

	Unit	LNG	Ammonia	Methanol	DME	Hydrogen
BOG during shipping	kg	25,491,045	8,570,317	202,082	6,277,619	37,838,929
Market price	$/GJ	5.93 [45]	28.2 [46]	16.3 [47]	15.06 [48]	12 [49]
Cost of BOG	$	7,346,468	4,495,302	67,560	2,732,233	54,488,058
Delivered quantity	kg	1,634,041,379	2,637,020,689	3,108,965,517	2,840,551,724	274,593,103
Delivered energy	GJ	79,414,411	49,048,584,827	61,868,413,793	82,091,944,827	32,951,172,413

Source [44]
[a]The ship operates for 350 days annually with a speed of 20 (knots)
[b]This consists of sailing time and 36 h turnaround for each trip
[c]The discounted cash flow (DCF) rate of 10% for a lifetime of 15 years
[d]Cost of HFO is 0.58 ($/kg) [56]
[e]Fuel needed to carry the full capacity is 2899 kg of HFO [53, 54]

Table 8.9 *Total production cost* of various energy carriers with literature comparison

	This study	Literature [50]
	$/GJ	$/GJ
LNG	5.86	5.76
Liquid ammonia	13.25	12.90
Methanol	8.33	9.05
DME	8.57	9.52
Liquid hydrogen	5.32	5.00

per unit of energy ($/GJ). The fluctuation in production costs arises from disparities in process complexity, energy consumption, and necessary infrastructure.

The production of LNG from natural gas entails merely gas pretreatment and liquefaction, rendering it comparatively economical. Nonetheless, methanol and DME incur higher production expenses compared to LNG and hydrogen, yet are still less expensive than ammonia. The manufacturing of methanol and DME necessitates supplementary processes, such as reforming and synthesis reactions, which elevate overall expenses. The boiling points of various energy carriers additionally affect production costs. DME and methanol, with boiling points of $-24\ °C$ and $64\ °C$, respectively, necessitate less cooling than ammonia, which has a boiling point of $-34\ °C$. The supplementary energy necessary to sustain ammonia in liquid state renders its manufacturing costlier.

In addition to the overall production cost, Fig. 8.7 presents a comprehensive cost analysis, delineating the contributions of capital cost, operational cost, and BOG cost for each energy carrier. This analysis elucidates the impact of various cost elements on total production expenditures and identifies avenues for cost optimization, especially in minimizing BOG losses during production and storage.

The production cost of liquid energy carriers comprises four principal components: investment cost, gas consumption cost, non-gas operational cost, and boil-off gas (BOG) cost. Every energy carrier possesses a distinct cost structure determined by its production methodology. In LNG production, the expenses are nearly equally divided between investment costs and gas usage costs, with each constituting approximately 30% of the overall production cost. The BOG cost is little, constituting about 0.18% of overall production expenditures, underscoring the necessity of optimal utilization of generated BOG throughout manufacturing.

In the case of liquid ammonia, non-gas operational expenses constitute roughly 52.7% of the overall production cost, and the remaining 47.3% is attributed to investment and gas consumption. The necessity of nitrogen synthesis in ammonia production contributes to operational costs. Methanol and DME exhibit a comparable cost structure, characterized by elevated non-gas operational expenses relative to other cost components. Conversely, hydrogen production exhibits the largest percentage of BOG costs, constituting 2.68% of the overall production expenses. The cost of gas utilization in hydrogen generation constitutes 47.3%, indicating that alternate

8.2 Economic Assessment

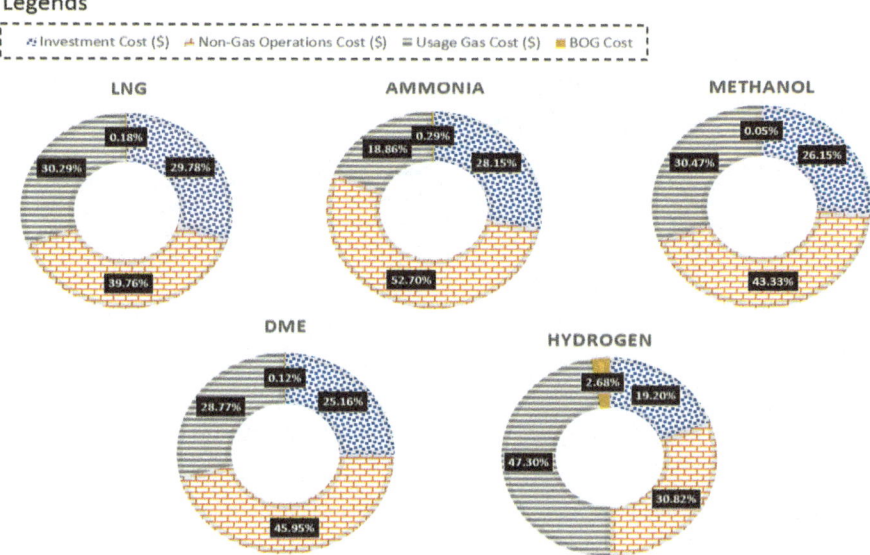

Fig. 8.7 Cost breakdown of producing various energy carriers (from [37], licensed under CC-BY 4.0)

production methods may be more economically viable for diminishing dependence on natural gas.

A sensitivity analysis is performed to assess the effect of varying natural gas prices from $1 to $4 per GJ on overall production costs, given their substantial influence on the production expenses of these energy carriers as presented in Fig. 8.8. With the escalation of natural gas prices, the production expenses of all five energy carriers correspondingly increase. When natural gas attains a price of 4 $/GJ, the manufacturing cost of ammonia escalates to 16 $/GJ, becoming it the most costly energy carrier. Conversely, when gas prices are low at 1 $/GJ, the production costs for methanol and DME are 7.06 $/GJ and 7.34 $/GJ, respectively.

In contrast, the production cost of LNG increases to $7.64 per GJ when gas prices attain $4 per GJ. This indicates that for nations with minimal natural gas production expenses, exporting natural gas as methanol and DME is more economically advantageous than exporting LNG. Furthermore, when natural gas prices decrease to 1 $/GJ, the production cost of hydrogen approximates 4.0 $/GJ, becoming methane-based hydrogen generation significantly competitive relative to alternative methods such as electrolysis, which tend to be more costly. This analysis emphasizes the significance of natural gas supply and pricing in ascertaining the most economical energy carrier for export and worldwide market competitiveness.

The maritime transportation expense for LNG, liquid ammonia, methanol, DME, and liquid hydrogen is calculated by summing the capital cost, operational cost, and BOG cost. The capital expenditure for LNG tankers exceeds that of ammonia tankers by 15% and that of methanol tankers by 40%, attributable to the requirement

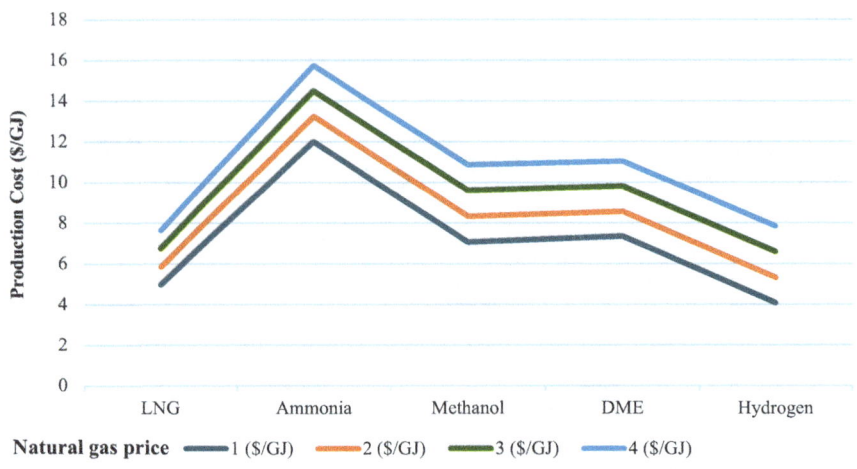

Fig. 8.8 Effects of changing the cost of natural gas on the total production cost of energy carriers

for high-quality insulated materials to sustain extremely low storage temperatures. Liquid hydrogen necessitates cryogenic storage at − 253 °C, resulting in a capital cost of $216,784,000 for a tanker with a capacity of 160,000 m^3. Conversely, DME and methanol tankers incur far reduced capital expenditures due to their ability to be kept as liquids at near-ambient temperatures, hence minimizing insulation and refrigeration needs.

Amid all energy carriers, ammonia incurs the highest operational costs for tankers. The operational cost of an ammonia tanker exceeds that of LNG tankers by 8% and surpasses that of hydrogen tankers by 20%. This is primarily attributable to ammonia's greater density and storage necessities, which elevate handling and operational costs. During transit, LNG incurs the most significant BOG losses, with a boil-off rate of 0.12%, resulting in a loss of 1,000,000 kg each trip. This issue can be resolved by implementing aboard BOG reliquefication systems, transitioning to alternate liquefied energy carriers, or improving insulating technology in shipping tanks. Notwithstanding these losses, LNG provides greater energy per trip than liquid ammonia, methanol, and hydrogen owing to its elevated energy content (heating value).

DME has the lowest overall transportation cost, which is 15% less than that of LNG. The whole transportation expense for liquid ammonia is $1.08 per GJ, comprising 45.6% tanker operations costs, 45.9% tanker capital costs, and 8.5% BOG costs. Transporting natural gas as liquid ammonia incurs more costs than LNG in terms of $/GJ due to ammonia's inferior energy density compared to LNG as shown in Fig. 8.9.

In terms of transportation cost per unit of stored energy ($/kg), liquid ammonia is more economical than LNG. The comprehensive transportation expense for ammonia is $0.02 per kilogram, whereas for LNG, it amounts to $0.038 per kilogram. Due to

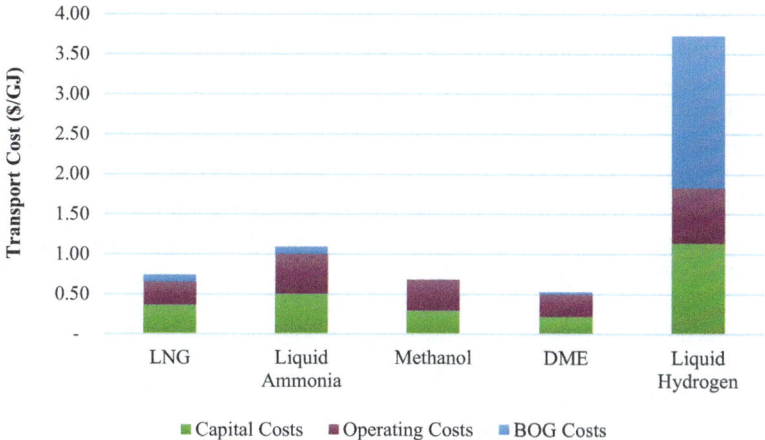

Fig. 8.9 Breakdown of total transportation costs of LNG, liquid ammonia, methanol, and DME in $/GJ

methanol's lower energy density (22 MJ/kg) relative to ammonia (22.5 MJ/kg) and LNG (54 MJ/kg), its transportation cost per unit mass is approximately 50% less than that of LNG and 15% less than that of ammonia as shown in Fig. 8.10. This analysis elucidates the trade-offs among various energy carriers concerning transportation costs, energy efficiency, and storage feasibility, rendering certain options more economically viable for long-distance transport based on market conditions and infrastructural capacities.

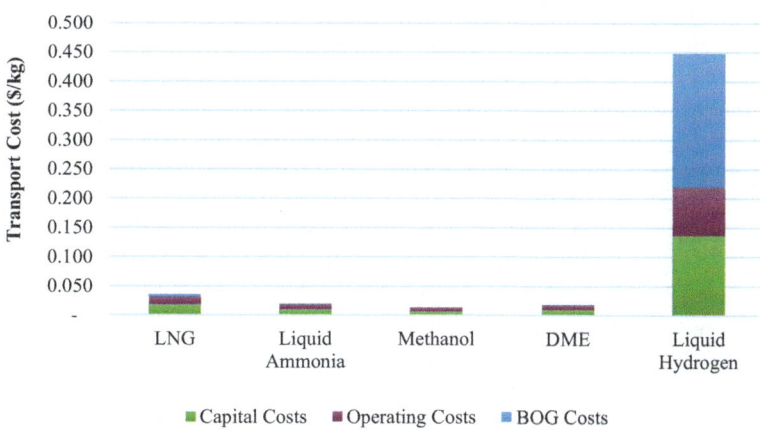

Fig. 8.10 Breakdown of total transportation costs of LNG, liquid ammonia, methanol, and DME in $/kg

Table 8.10 presents the total maritime shipping costs for LNG, liquid ammonia, methanol, DME, and liquid hydrogen in $/m^3. The data indicates a definitive correlation between energy carrier density and transportation cost per cubic meter; as the density of an energy carrier rises, its transportation cost per cubic meter diminishes.

Methanol possesses the maximum density at 805 kg/m^3, rendering it the most economical choice for energy transportation, with a cost of roughly \$10.89/m^3. Conversely, liquid hydrogen incurs the largest transportation expenses, almost 60 times more than those of methanol. This is mostly attributable to the exceedingly low density of liquid hydrogen (71.1 kg/m^3), necessitating specialized cryogenic storage at $-253\,°C$, which substantially elevates both capital and operational expenditures.

Due to its high density and very low transportation cost per cubic meter, methanol is the most economical energy carrier for large-scale energy transmission. DME and ammonia present competitive transportation costs, rendering them feasible options based on supply chain needs. Nonetheless, liquid hydrogen is the costliest alternative owing to its distinct storage difficulties and associated infrastructure expenses.

Figure 8.11 demonstrates the effect of alterations in ship capacity on ocean transportation expenses for different liquid energy carriers. Typically, as vessel capacity expands, transportation expenses diminish owing to economies of scale. For example, as the ship's capacity escalates from 100,000 to 260,000 m^3, the transportation cost of LNG diminishes by almost 12.5%. Vessels of greater size may transport increased cargo per voyage, hence reducing the cost per unit of energy conveyed.

The transportation cost of LNG for a 250,000 m^3 capacity ship is \$0.7/GJ, but methanol and DME incur lesser expenses of \$0.63/GJ and \$0.5/GJ, respectively. The elevated expense of LNG transportation is primarily attributable to BOG generation. As the dimensions of a vessel expand, the surface area correspondingly rises, resulting in enhanced heat absorption and elevated BOG losses. This renders LNG less economically viable in comparison to DME and methanol, which exhibit reduced BOG rates and necessitate less rigorous insulation.

DME and methanol present a more cost-effective option for the long-distance transportation of natural gas-derived energy, owing to their reduced transportation expenses and negligible BOG losses, in contrast to LNG. This indicates that DME and methanol may be optimal selections for large-scale shipping operations for enhancing cost-effectiveness in energy transport.

Figure 8.12 depicts the aggregate transportation expenses of natural gas-derived energy from Qatar to principal energy-importing regions: India, China, and Japan. Since the transportation distance varies, the cost of transporting energy increases with longer shipping routes. Of the three destinations, India is nearest to Qatar, leading to reduced transportation expenses in contrast to the more distant routes such as China and Japan.

Table 8.10 Ocean transportation cost of various energy carriers in $/m^3

	LNG	Ammonia	Methanol	DME	Hydrogen
Cost of ocean transportation (\$/m^3)	15.3	13.87	10.89	11.36	27.66

8.2 Economic Assessment

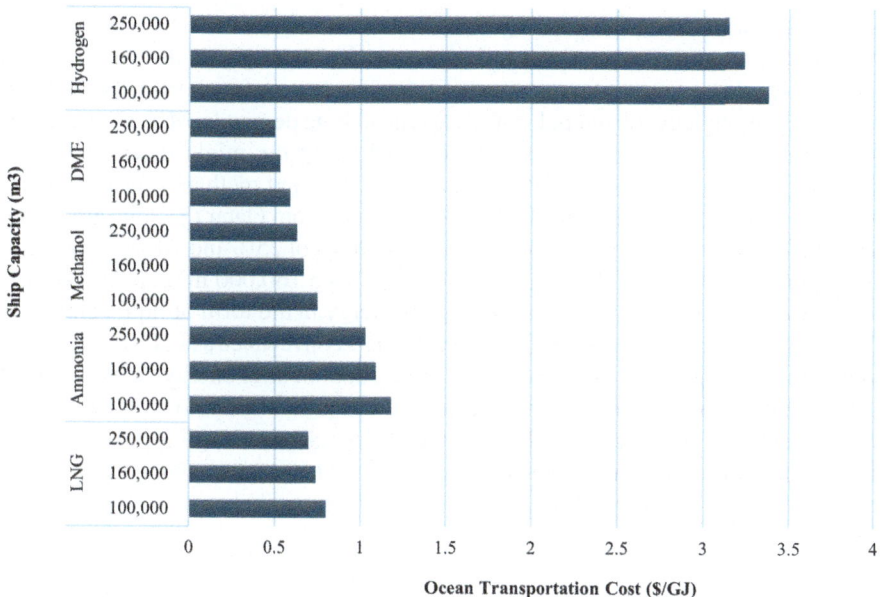

Fig. 8.11 Sensitivity analysis on ocean transportation cost for various energy carriers based on ship capacity

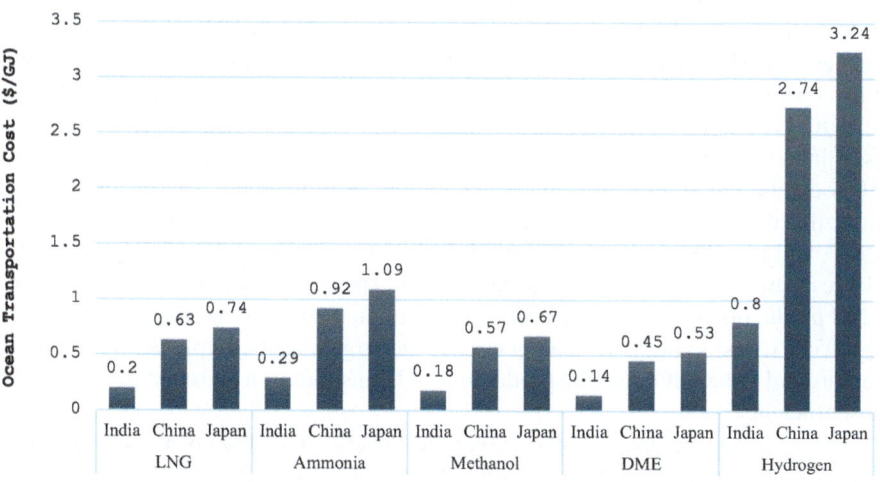

Fig. 8.12 Costs for transporting natural gas energy via various energy carriers from Qatar to India, Chain, and Japan

Shipping liquid hydrogen from Qatar to India incurs a cost of $0.80 per gigajoule, whereas shipping LNG from Qatar to Japan costs $0.74 per gigajoule. This suggests that liquid hydrogen may serve as a cost-efficient alternative for short-distance transportation, although its substantial storage and insulating demands. Moreover, nations possessing substantial natural gas reserves, such as Qatar, might diversify energy exports by converting a portion of their gas into DME and methanol for transportation. The transportation costs for DME and methanol from Qatar to China are $0.45/GJ and $0.57/GJ, respectively, representing reductions of 25% and 10% compared to LNG transport costs over the same distance utilizing a 160,000 m^3 capacity vessel.

Moreover, the transportation of natural gas energy in the form of liquid ammonia from Qatar to India is 53% more economical than its conveyance as LNG to China. As both liquid hydrogen and liquid ammonia are carbon–neutral energy carriers devoid of direct greenhouse gas emissions, the conversion of natural gas into these substances could bolster a nation's economy by facilitating low-carbon energy exports. By transitioning to DME, methanol, liquid ammonia, and hydrogen, gas-abundant countries can enhance transportation efficiency and facilitate the global energy move towards cleaner fuels.

8.2.5.4 Conclusion

This study provides a thorough economic analysis of the production and transportation of energy carriers produced from natural gas, including LNG, liquid ammonia, methanol, DME, and liquid hydrogen. The assessment examines capital expenditures, operational expenses, and boil-off gas (BOG) costs, emphasizing the influence of BOG generation on overall supply chain costs. This analysis include BOG losses as a financial factor, in contrast to standard cost evaluations that remove them, as BOG can be flared, used for power generation, or reliquefied. The results demonstrate that the production cost per unit of energy for LNG, liquid ammonia, methanol, DME, and liquid hydrogen is $5.86/GJ, $13.25/GJ, $8.33/GJ, $8.57/GJ, and $5.32/GJ, respectively. Among these carriers, BOG expenses constitute merely 0.18% of LNG production costs, whereas BOG losses in methanol production are far smaller at 0.05%. Considering that LNG generates the largest BOG, implementing ways to capture and utilize BOG during production and transportation is crucial for enhancing cost efficiency.

The research underscores the economic benefit of methanol as an alternative energy carrier in areas with little natural gas production expenses. At a natural gas price of 1 $/GJ, the methanol production cost is roughly 7.34 $/GJ, but the LNG production cost escalates to 7.64 $/GJ when gas prices increase to 4 $/GJ. This indicates that for gas-abundant countries, exporting natural gas as methanol may be more economically advantageous than exporting LNG, especially during periods of fluctuating natural gas prices.

Capital expenses for ocean transportation exhibit considerable variation among different energy carriers. LNG tankers are 15% and 40% more costly than ammonia and methanol tankers, respectively, due to stringent temperature requirements and elevated insulation demands. Likewise, liquid hydrogen necessitates specialized cryogenic storage, rendering it the most capital-intensive transportation alternative. Furthermore, LNG experiences a significant BOG rate of 0.12% per trip, resulting in 1,000,000 kg of LNG losses per journey for a vessel with a capacity of 160,000 m^3. The overall transportation cost of LNG is assessed at $0.038 per kilogram, but liquid ammonia incurs a lesser cost of $0.02 per kilogram. Due to methanol's lower heating value (19.9 MJ/kg) relative to LNG (48.6 MJ/kg), its transportation cost per unit of energy is approximately 50% less than that of LNG.

The vessel's capacity significantly contributes to the reduction of transportation expenses. As the tanker capacity escalates from 100,000 to 260,000 m^3, the transportation expenditure for LNG diminishes by roughly 12.5%. For a ship with a capacity of 250,000 m^3, the transportation costs for LNG are 0.7 $/GJ, whereas the costs for methanol and DME are lower, at 0.63 $/GJ and 0.5 $/GJ, respectively. The elevated expenses of LNG transportation are ascribed to BOG generation, which escalates with the enlargement of ship size and surface area. This discovery indicates that DME and methanol are more economical alternatives for large-scale energy transportation, especially in contexts where minimizing BOG losses is essential.

The research additionally investigates the impact of transit distance on energy carrier expenses. As anticipated, reduced transit distances yield decreased costs. Transporting liquid hydrogen from Qatar to India incurs a cost of 0.8 $/GJ, while LNG shipping from Qatar to Japan costs 0.74 $/GJ. This suggests that liquid hydrogen may be economically viable for short-distance transportation, contingent upon advancements in BOG reduction. Furthermore, nations possessing substantial natural gas reserves may gain advantages by diversifying energy exports beyond LNG through the production and exportation of DME and methanol. The transportation costs for DME and methanol from Qatar to China are $0.45/GJ and $0.57/GJ, respectively, rendering them 25% and 10% less expensive than LNG shipping over the same distance with a 160,000 m^3 vessel. Moreover, the transportation of liquid ammonia from Qatar to India is 53% less expensive than the shipment of LNG to China, rendering ammonia a viable alternative for energy transport.

For countries abundant in natural gas, such as Qatar, the findings underscore essential measures to optimize economic benefits from energy exports. Besides exporting LNG, emphasizing the large-scale production and transportation of DME and methanol may provide a more economical alternative. Moreover, the conversion of natural gas into liquid hydrogen and liquid ammonia for short-distance exports presents a potential to penetrate carbon-free energy markets. The utilization of liquid hydrogen and ammonia, which emit no direct greenhouse gases, could enhance a nation's standing in the clean energy sector while also facilitating economical energy transportation. By utilizing various energy carriers and refining transportation systems, natural gas-producing countries can improve energy security, augment economic returns, and aid in the worldwide shift towards sustainable energy solutions.

8.3 Life Cycle Assessment

As the global energy sector transitions to sustainability, comprehending the environmental effects of energy carriers is essential. Life cycle assessment (LCA) is a methodical approach to assess the environmental attributes and potential effects linked to a product, process, or service across its entire life cycle. This section presents an LCA of liquefied energy carriers, including LNG, liquefied hydrogen, ammonia, methanol, DME, and formic acid, concentrating on the emissions linked to their production, storage, and transportation. The evaluation seeks to educate students and professionals regarding the environmental compromises associated with the use of these carriers.

8.3.1 Emissions Associated with Production

The generation of energy carriers is a multifaceted, multiphase process that significantly influences the environmental ramifications of contemporary energy systems. As global energy demands escalate, comprehending and regulating the greenhouse gas emissions linked to the production of these carriers becomes increasingly critical. Every energy carrier—ranging from LNG and liquefied hydrogen to ammonia, methanol, DME, and formic acid—is generated through specific procedures that encompass extraction, conversion, and occasionally further purification stages. These processes necessitate substantial energy inputs and produce emissions at multiple stages, thereby affecting the overall sustainability of the energy carrier.

This section offers a comprehensive examination of the emissions associated with the generation of these energy carriers. We commence by analyzing the upstream processes, including natural gas extraction and hydrogen production, which frequently entail the emission of significant greenhouse gases. We subsequently examine how additional processing stages, such as liquefaction and catalytic conversion, further influence the emissions profile. The analysis indicates that the environmental impact of each production pathway is influenced by a mix of factors, including the energy source utilized, the efficiency of conversion technologies, and the selection of feedstocks. Additionally, we examine novel production techniques that include renewable energy sources and cutting-edge technologies to reduce emissions, presenting prospective avenues for more sustainable production practices.

This part seeks to enhance comprehension of the emissions linked to production, so enabling students and professionals to critically evaluate the trade-offs between energy efficiency and environmental sustainability. The discourse establishes a framework for assessing enhancements in production methodologies, directing subsequent research and policy initiatives aimed at diminishing the carbon footprint of energy carriers while preserving their viability within the global energy context.

8.3.1.1 LNG

The production of LNG commences with the extraction of natural gas from subterranean reserves. Fugitive methane emissions arise during extraction, which is particularly critical due to methane's status as a strong greenhouse gas. Upon extraction, natural gas is processed to eliminate impurities including water, sulfur compounds, and other pollutants. Each of these steps contributes to the total emissions linked to LNG production.

The next liquefaction process—necessary for chilling the gas to cryogenic temperatures—involves compressing and cooling the gas via several steps. This energy-intensive phase generally depends on combustion mechanisms that produce supplementary carbon dioxide. The total environmental effect of LNG production is influenced by both direct emissions from fuel burning and indirect emissions from upstream operations. This intricate interaction of elements highlights the necessity of enhancing extraction efficiency, minimizing methane emissions, and refining liquefaction technology to reduce the greenhouse gas impact of LNG.

8.3.1.2 Liquefied Hydrogen

The production of liquefied hydrogen is marked by significant energy requirements and resultant emissions. Hydrogen is often generated using steam methane reforming (SMR), a method that is linked to substantial carbon dioxide emissions resulting from the transformation of natural gas into hydrogen. The ensuing liquefaction phase exacerbates the environmental impact, as it necessitates cooling hydrogen to exceedingly low temperatures—a procedure that demands significant energy expenditure.

The emissions profile of liquefied hydrogen is significantly influenced by the energy source utilized throughout the process. The environmental footprint is significantly greater when fossil fuels are utilized for essential cooling and processing. In contrast, utilizing renewable energy for hydrogen synthesis through electrolysis and subsequent liquefaction can significantly diminish emissions. The sustainability of liquefied hydrogen depends on the incorporation of green energy solutions across the production chain.

8.3.1.3 Ammonia

Ammonia is predominantly produced via the Haber–Bosch process, which is both energy-demanding and significantly dependent on natural gas. The synthesis of ammonia entails the transformation of nitrogen and hydrogen, with hydrogen generally obtained by natural gas reforming. This phase of hydrogen production significantly contributes to overall emissions due to substantial fossil fuel combustion and the resultant CO_2 emissions.

Alternative methods employing green hydrogen generated by water electrolysis have the potential to diminish the carbon footprint of ammonia production. Nonetheless, these technologies necessitate significant energy inputs and sophisticated equipment to remain competitive, creating a trade-off between diminished emissions and heightened energy demand. The emissions profile of ammonia production reflects the energy demands of the Haber–Bosch process and the selection of hydrogen feedstock.

8.3.1.4 Methanol

The synthesis of methanol can occur via many industrial methods, each possessing unique environmental consequences. The production of methanol from natural gas entails reforming the gas to generate synthesis gas, followed by catalytic conversion. This approach, although more efficient than coal-based methods, still generates considerable emissions due to the energy needed for reforming and conversion.

Conversely, methanol production from coal demonstrates a more significant environmental impact, mostly due to the elevated carbon intensity of coal. Novel techniques employing captured carbon dioxide and green hydrogen are developing as viable alternatives to traditional technologies. These methodologies possess the capacity to attain a carbon–neutral or perhaps carbon-negative production cycle; nevertheless, they are presently in the developmental phase and encounter obstacles concerning scalability and economic viability. The environmental impact of methanol production is primarily influenced by the selection of feedstock and the energy source employed in the conversion operations.

8.3.1.5 DME

DME can be synthesized through the dehydration of methanol or by direct synthesis from synthesis gas. The environmental consequences of DME manufacturing are intricately associated with the source of its fuel. The production process derived from natural gas often exhibits a greater carbon footprint because to the intrinsic emissions linked to natural gas reforming and methanol synthesis.

Conversely, the production of DME using sustainable methanol or biomass-derived synthesis gas can substantially reduce the associated emissions. In these instances, the environmental performance of DME is significantly enhanced, potentially leading to a production process with minimal or even negative net emissions. This heterogeneity underscores the significance of feedstock selection and process integration in ascertaining the overall sustainability of DME production.

8.3.1.6 Formic Acid

Conventional synthesis of formic acid entails the hydrolysis of intermediates like methyl formate, a method that is energy-intensive and dependent on traditional feedstocks. The resultant emissions mostly depend on the energy needed to synthesize the precursor compounds and the efficacy of the hydrolysis process.

Innovative production techniques, such as CO_2 hydrogenation utilizing green hydrogen, are being advanced as more sustainable alternatives. These innovative strategies seek to utilize carbon dioxide extracted from the atmosphere or industrial processes, so presenting the possibility of a carbon–neutral or even carbon-negative manufacturing method. Although these approaches are still being refined, they signify a crucial advancement in mitigating the environmental impact of formic acid production.

8.3.2 Emissions Associated with Storage

The storage phase of energy carriers is a crucial, yet frequently overlooked, stage that can substantially affect the total environmental impact of an energy system. In contrast to production and transportation, which primarily emphasize energy conversion and transfer efficiencies, storage presents distinct challenges due to the necessity of sustaining particular physical conditions—such as cryogenic temperatures or pressurized environments—that inherently demand ongoing energy input and meticulous management. These conditions can lead to multiple forms of emissions, resulting not only from the direct loss of the stored chemical but also from ancillary operations such as boil-off gas (BOG) management and leak prevention.

In cryogenic storage systems, such as those utilized for LNG and liquefied hydrogen, the preservation of exceedingly low temperatures is crucial for storing these substances in a liquid condition. Nevertheless, despite employing cutting-edge insulation and active cooling systems, a portion of the stored liquid unavoidably evaporates, resulting in boil-off gas. If this gas is not effectively reliquefied or utilized productively, it may be emitted into the atmosphere, so contributing to greenhouse gas emissions. The issues are exacerbated by the vaporized components, which are frequently abundant in powerful greenhouse gasses and exert a disproportionately significant influence on global warming relative to their liquid equivalents.

Alternative energy carriers, such ammonia, methanol, DME, and formic acid, have less stringent storage needs; yet they still pose distinct environmental challenges. Ammonia must be kept in a manner that prevents leakage, as even modest emissions can have significant ecological and health repercussions. Methanol and formic acid, when stored under ambient conditions, are prone to evaporation and the consequent emission of volatile organic compounds (VOCs), which may contribute to the development of ground-level ozone. DME, maintained at moderate pressure, is susceptible to leakage; nevertheless, its environmental impact is alleviated by its comparatively brief atmospheric lifespan and swift breakdown.

This section examines the several origins of emissions during storage, highlighting the qualitative factors contributing to their emergence and the engineering problems associated with their mitigation. This discussion seeks to furnish students with a thorough comprehension of the environmental implications associated with the storage of each energy carrier, emphasizing the necessity of implementing effective storage technologies and methods. This analysis highlights the necessity for comprehensive solutions that tackle emissions throughout production, transportation, and storage, ultimately fostering a more sustainable and efficient energy infrastructure.

8.3.2.1 LNG

LNG storage entails preserving natural gas at significantly low temperatures to retain its liquid form. Nonetheless, because to unavoidable heat intrusion, a segment of the LNG progressively vaporizes, resulting in boil-off gas (BOG). This vapor, predominantly consisting of methane, is an environmental concern if inadequately controlled. While BOG is frequently reliquefied or utilized as a fuel for power generation, any uncontrolled venting or leakage exacerbates greenhouse gas emissions due to methane's significantly elevated warming potential relative to other gases.

8.3.2.2 Liquefied Hydrogen

Storing liquefied hydrogen necessitates the maintenance of cryogenic temperatures that are inferior to those required for LNG. At these exceedingly low temperatures, some hydrogen inevitably vaporizes, resulting in boil-off losses. Beyond the immediate loss of hydrogen, vented hydrogen can indirectly impact atmospheric chemistry by altering the longevity and dynamics of other greenhouse gases. While hydrogen possesses a weaker warming effect than methane, the entire storage process is energy-intensive, necessitating meticulous control to mitigate environmental repercussions.

8.3.2.3 Ammonia

The storage of ammonia has specific issues owing to its chemical characteristics. It may be stored either under pressured settings at ambient temperature or in refrigerated surroundings at reduced temperatures. In both instances, there exists a possibility of leakage, which not only signifies a loss of goods but also presents considerable environmental and health risks due to the toxicity of ammonia. Upon atmospheric release, ammonia can facilitate the creation of particulate matter and may induce nutrient enrichment in ecosystems, potentially resulting in problems such as eutrophication. Consequently, effective containment and leak prevention strategies are required.

8.3.2.4 Methanol

Methanol, as a volatile liquid, is prone to evaporation during storage. The volatilized methanol emits volatile organic compounds (VOCs) into the atmosphere, which can contribute to the development of ground-level ozone. The creation of ozone is detrimental to human health and plant life. Nonetheless, methanol possesses the benefit of fast biodegradation, which aids in reducing its prolonged environmental effects. Optimal storage methods and efficient vapor recovery technologies are essential for reducing these emissions.

8.3.2.5 DME

The storage of DME generally requires pressurized containers akin to those utilized for liquefied petroleum gases. While pressured storage conditions may result in modest emissions from leaks, DME is a volatile organic molecule with a typically brief atmospheric lifespan. With proper containment and routine maintenance, the environmental impact of DME storage is negligible, as any emissions are swiftly diffused and decomposed in the atmosphere.

8.3.2.6 Formic Acid

Formic acid is normally stored at ambient conditions, leading to minimal direct emissions during storage. Nonetheless, its corrosive properties mean that any inadvertent discharge can result in localized ecological harm. Although formic acid does not remain in the environment for prolonged durations due to its rapid biodegradation, it is crucial to minimize any accidental releases to mitigate short-term ecological damage. Consequently, efficient storage systems and monitoring techniques are essential to minimize emissions.

8.3.3 Emissions Associated with Transportation

The transportation phase of energy carriers is a vital aspect of the total life cycle, substantially affecting their environmental effect and sustainability. Transportation encompasses the movement of products from production sites to end-use markets and significantly influences the net greenhouse gas emissions linked to each energy carrier. This phase includes several operational techniques, such as the utilization of specialized containers or pipelines, fuel burning during transportation, and the management of product losses like boil-off and leakage. Each element influences the total emission profile, necessitating a comprehension of the intricacies of transportation-related emissions within the larger energy system.

In an academic environment, studying transportation emissions entails investigating the relationship between the physical characteristics of energy carriers and the technological systems utilized for their movement. Carriers such as LNG and liquid hydrogen, kept under cryogenic settings, encounter intrinsic problems including vaporization losses that necessitate advanced management solutions. Conversely, energy carriers like ammonia, methanol, DME, and formic acid are conveyed under conditions that typically entail reduced temperature and operational strains; yet, they represent distinct risks associated with leakage or unintentional releases. These discrepancies highlight the necessity of formulating customized mitigation methods that consider the unique attributes of each carrier.

This analysis encompasses more than just the direct emissions produced during travel. It includes a comprehensive perspective of the transportation infrastructure, incorporating the energy inputs necessary for compression, refrigeration, and the upkeep of safe working conditions. If not adjusted, these operating activities can result in indirect emissions that may negate the environmental advantages gained in prior manufacturing or storage phases. This section critically evaluates these factors to furnish students with a thorough understanding of how transportation practices influence the total emissions of energy carriers and underscores the necessity for ongoing advancements in technology and management practices to promote a more sustainable energy future.

This section discusses the issues in lowering transportation emissions and identifies potential avenues for innovation. Integrating new materials, enhanced vessel design, and optimized operational protocols can render the transfer of energy carriers more environmentally sustainable. An integrated strategy is crucial for policymakers, engineers, and industry stakeholders collaborating to reduce the carbon footprint of energy systems while fulfilling global energy requirements.

8.3.3.1 LNG

LNG is predominantly carried by specialized maritime tankers fitted with technologies to handle boil-off gas (BOG). These systems may occasionally utilize the vaporized gas as a fuel source, hence diminishing dependence on supplementary fossil fuels. Nonetheless, the process is not completely devoid of emissions. During loading, unloading, and engine running, there exists a potential for the release of both carbon dioxide and methane. Furthermore, the transportation of LNG through pipelines entails the risk of gas leakage during the compression and transmission phases, which exacerbates its environmental impact. The total emissions from LNG transportation are significantly influenced by the efficiency of operational techniques and the efficacy of leak mitigation strategies.

8.3.3.2 Liquefied Hydrogen

Transporting liquefied hydrogen has distinct issues due to the severe cryogenic conditions necessary to preserve its liquid state. Refrigeration systems must function continuously during transport, and if the electricity powering these systems is derived from fossil fuels, indirect emissions may be substantial. Moreover, hydrogen experiences boil-off losses, and although its direct global warming potential is lower than that of methane, the indirect impacts related to its management can be significant, particularly over extended distances. The environmental impact of liquid hydrogen transportation is significantly influenced by the energy sources and technology utilized during travel.

8.3.3.3 Ammonia

Ammonia utilizes a well-established transportation infrastructure that employs technologies akin to those found in the chemical sector. Ammonia emissions during transportation by tanker or pipeline are comparatively little under standard operating circumstances. Nevertheless, the possibility of inadvertent leaks continues to be a considerable concern due to ammonia's toxicity and its ability to create serious ecological and health repercussions. The emphasis in ammonia transportation is on stringent safety regulations and leak prevention measures. The efficacy of these methods is crucial in reducing the environmental impact during transportation.

8.3.3.4 Methanol

Methanol is generally conveyed by standard tankers or pipes intended for liquid chemicals. The transportation of methanol typically generates minimal emissions, chiefly from the gasoline consumed by the transport vehicles. Methanol is volatile, and spills or evaporative losses are often less persistent in the environment due to its quick biodegradation. This intrinsic characteristic of methanol aids in reducing its prolonged environmental effects; nonetheless, precautions must be observed to control any inadvertent discharges during handling and transportation.

8.3.3.5 DME

The transport of DME is conducted similarly to that of liquefied petroleum gases, utilizing pressurized containers to guarantee safe travel. The procedures required to sustain the requisite pressure and transmit the substance mostly generate emissions through the combustion of gasoline in transportation vehicles. DME is marked by low toxicity and a brief atmospheric duration, indicating that unintentional discharges have a relatively limited environmental impact. Effective management during loading and unloading might further diminish the emissions linked to its transit.

8.3.3.6 Formic Acid

Formic acid is conveyed by standard chemical tanker systems at ambient conditions. The emissions associated with its transportation primarily stem from the fuel consumption of the transport vehicles, rather than the formic acid itself. Unintentional spills of formic acid may result in limited ecological issues due to its acidic properties. Nonetheless, owing to its swift degradation in the environment, the prolonged environmental impacts of such spills are typically constrained. Efficient spill management and containment solutions are crucial to mitigate any negative effects.

8.3.4 Comparison and Analysis

An exhaustive evaluation of the environmental effects of energy carriers necessitates a holistic perspective on their production, storage, and transit phases any phase distinctly influences the overall greenhouse gas (GHG) emissions profile, and comprehending these variations is crucial for assessing the sustainability of any carrier. This section compares and analyzes the emissions linked to the production, storage, and transportation of several energy carriers, including LNG, liquefied hydrogen, ammonia, methanol, DME, and formic acid, emphasizing the strengths and drawbacks of each process.

The intrinsic characteristics of each energy carrier dictate the level of emissions throughout the production phase. LNG production encompasses the extraction of natural gas, its processing, and subsequent liquefaction via an energy-intensive cooling procedure. This phase is characterized by considerable problems, including fugitive methane emissions during extraction and the substantial energy demands for cryogenic cooling. Conversely, the manufacture of liquid hydrogen necessitates even more harsh circumstances, as hydrogen must be cooled to substantially lower temperatures. The hydrogen manufacturing methods, including steam methane reforming and water electrolysis, have unique emission profiles. Traditional methods often generate elevated CO_2 emissions, whereas innovative green approaches, although demanding significant energy inputs, have the possibility of markedly diminished emissions when fueled by renewable sources.

Upon analyzing storage, the distinctions in physical state and storage circumstances become increasingly evident. LNG and liquefied hydrogen necessitate cryogenic storage, which essentially entails the management of boil-off gas (BOG) losses. The management of these losses—through reliquefaction, fuel utilization, or venting—significantly impacts overall emissions. LNG, despite being associated with boil-off gas (BOG) methane emissions, advantages from more advanced and sophisticated storage methods than liquefied hydrogen, which encounters elevated boil-off rates and increased energy requirements for active cooling. Conversely, carriers like ammonia, methanol, DME, and formic acid are maintained in less severe circumstances, such as ambient or weakly regulated environments. Consequently, their storage emissions are typically reduced; yet, each entails distinct obstacles. Ammonia necessitates strong containment to avert leakage owing to its toxicity, whereas methanol and formic acid, although less energy-intensive to store, require meticulous management of volatile organic compounds (VOCs) or corrosive emissions.

Transportation further distinguishes the environmental characteristics of various energy carriers. LNG is generally conveyed using specialized maritime tankers or pipelines designed to manage boil-off gas and any methane emissions during loading, unloading, and compression processes. These procedures result in both direct CO_2 emissions from fuel burning and indirect emissions from operational inefficiencies. Likewise, liquid hydrogen encounters substantial transportation difficulties owing to the constant refrigeration necessary to preserve its cryogenic condition. The resultant energy requirements and possible boil-off losses augment the overall emissions burden, especially when fossil fuel-based electricity is utilized in the cooling process. In contrast, ammonia, methanol, DME, and formic acid advantage from established and less energy-demanding transportation systems. These carriers are generally transported using conventional chemical tanker systems or pipelines, which, when efficiently handled, yield relatively lower operational emissions. Nevertheless, in these instances, the risk of inadvertent releases—whether by leakage or spills—must be controlled to alleviate localized environmental consequences.

The total emissions linked to energy carriers arise from a complicated interaction among the production, storage, and transportation phases. Carriers such as LNG and liquefied hydrogen typically exhibit elevated emissions profiles, principally due to the substantial energy requirements of their cryogenic processes and the difficulties in minimizing boil-off losses. Conversely, ammonia, methanol, DME, and formic acid demonstrate reduced emissions during storage and transit; nonetheless, their production phases still substantially impact their entire environmental footprints. This comparative analysis highlights the necessity of a comprehensive strategy for emission reduction, wherein advancements in feedstock selection, process efficiency, and the incorporation of renewable energy sources at all stages can collectively improve the sustainability of the entire energy supply chain. These insights are essential for policymakers, engineers, and industry stakeholders in directing technological advances and operational practices that minimize the carbon footprint of energy systems while addressing the increasing global energy demand.

8.3.5 Case Study 3: Life Cycle Assessment of Energy Carriers' Entire Supply Chain

This case study is adopted from [59]. This study is to conduct a thorough life cycle assessment (LCA) to measure greenhouse gas (GHG) emissions throughout the complete supply chain of several energy carriers generated in Qatar, including LNG, methanol, DME, liquid hydrogen, and liquid ammonia.

8.3.5.1 Description

The study offers a comparative evaluation of greenhouse gas emissions from energy carriers originating from natural gas and renewable sources, examining the environmental consequences of their production, long-distance international transport, and ultimate use in internal combustion engines. This study assesses emissions per unit mass and per unit energy, elucidating the disparities in environmental footprints associated with the transportation of various energy carriers over short distances of 5000 nautical miles and long lengths of 20,000 nautical miles.

The study primarily examines liquid hydrogen, a nascent energy transporter in the shift to sustainable energy. The study evaluates different hydrogen generation methods, such as solar PV and wind energy-driven water electrolysis, steam methane reforming (SMR) with and without carbon capture and storage (CCS), and transportation via heavy fuel oil (HFO)-powered ocean ships. The study also analyzes nitrogen oxides (NO_x) and sulfur oxides (SO_x) emissions during the manufacturing, transportation, and consumption of these energy carriers, due to their considerable influence on air quality and environmental health. The study includes a sensitivity analysis to assess the complete life cycle greenhouse gas emissions of energy carriers derived from renewable sources including solar photovoltaic and biomass, in addition to natural gas, over transportation lengths ranging from 0 to 20,000 nautical miles. This analysis elucidates the variation of emissions contingent upon manufacturing methodologies and transportation logistics.

The study evaluates the impact of transitioning from natural gas-based production to renewable energy sources, including solar PV, wind, and biomass, on the reduction of life cycle greenhouse gas emissions. Losses from boil-off gas (BOG) produced during production and transportation are also taken into account, as capturing and utilizing BOG can further improve sustainability. This study's findings provide essential insights for policymakers, energy producers, and environmental researchers aiming to enhance energy production and transportation techniques while reducing emissions. The findings underscore prospects for shifting to greener energy sources, guaranteeing economic feasibility while enhancing energy security and sustainability.

This analysis is particularly pertinent for Qatar and other nations abundant in natural gas, since it illustrates how exports of alternative energy carriers, such methanol, DME, and liquid hydrogen, can provide lower-emission substitutes for conventional LNG exports. Moreover, evaluating the effects of carbon capture technologies and renewable hydrogen production might inform future investments in sustainable energy initiatives. This study assesses the environmental impact of different energy carriers, offering a comprehensive understanding of energy production, transportation, and utilization that coincides with global climate objectives and facilitates the transition to a low-carbon economy.

8.3.5.2 Methodology

LCA is a methodology employed to evaluate the environmental impacts of products, processes, or services from a cradle-to-grave viewpoint [60]. The LCA research has four stages: aim and scope definition, life cycle inventory (LCI), life cycle impact assessment (LCIA), and interpretation. The methodological flowchart utilized in this work is illustrated in Fig. 8.13 [61].

Defining the purpose and scope in an LCA is crucial, as it delineates the system boundaries and functional unit, forming the foundation for evaluating various energy

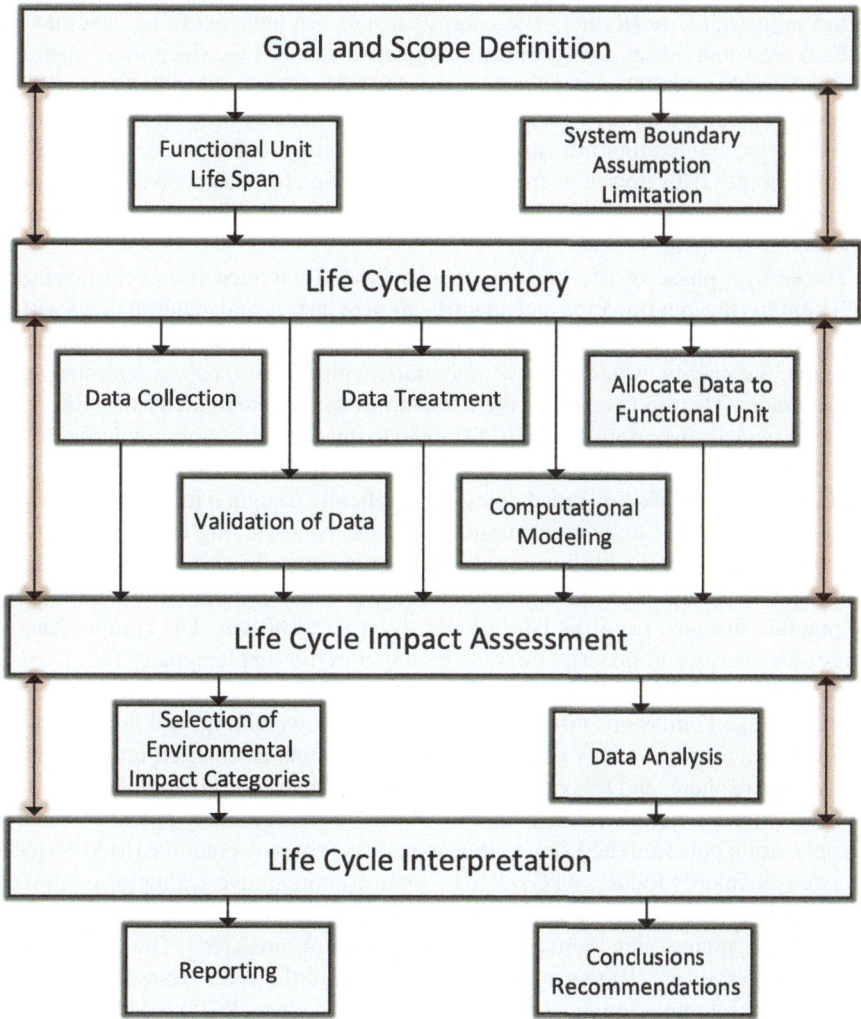

Fig. 8.13 Life cycle assessment framework and methodology flowchart [59]

carriers. The functional unit denotes a standardized metric facilitating equitable comparisons among diverse items. This study defines the functional unit as the production and transportation of 1 kg and 1 MJ of energy carrier over varying distances for use in an internal combustion engine (ICE). The project is to evaluate the life cycle environmental performance of five energy carriers—LNG, methanol, DME, liquid hydrogen, and liquid ammonia—derived from either natural gas or renewable energy sources. The system boundary includes all phases from raw material extraction, feedstock transportation, production, storage, and international transfer to final utilization.

The study is undertaken in Qatar, where natural gas, the primary energy source, is collected from the North Field and transferred by an 80 km pipeline to Ras Laffan Industrial City (RLIC). The energy carriers are believed to be generated at RLIC, stored, and subsequently transmitted to diverse worldwide locations, spanning lengths from 0 to 20,000 nautical miles (nmi) [62]. The gasoline production facilities comprise machinery for feedstock conversion, equipment for product transportation, storage tanks, and loading infrastructure [23]. The study includes a sensitivity analysis to evaluate GHG emissions from fuel transportation by ocean tankers, examining two propulsion fuel alternatives: heavy fuel oil (HFO) and LNG as a substitute bunker fuel.

The second phase of life cycle assessment (LCA), termed life cycle inventory (LCI), entails the identification and quantification of energy and material flows within the system and their environmental interactions. This encompasses the examination of inputs like energy, water, and raw materials, with outputs such as emissions and contaminants. This study employs the Greenhouse Gases, Regulated Emissions, and Energy Use in Transportation (GREET) model to quantify emissions, including CO_2, CH_4, N_2O, and other pollutants produced throughout the transportation life cycle. The GREET model has preloaded datasets specifically designed for North American energy systems. This study concentrates on Qatar, necessitating an updating of the inventory data to align with local production procedures. The alterations include of utilizing solely natural gas as a feedstock for electricity generation and modifying the pipeline distance from the North Field to RLIC to 80 km. The comprehensive dataset for energy and material fluxes is included in the supplementary file.

The third stage, life cycle impact assessment (LCIA), assesses the environmental repercussions of emissions through the classification, weighting, and normalization of pollution categories. This phase quantifies emissions discharged into the atmosphere, hydrosphere, and lithosphere, encompassing metrics such as global warming potential (GWP), acidification potential (AP), ozone depletion potential (ODP), eutrophication potential (EP), and marine aquatic ecotoxicity potential (MAEP) [63]. This study primarily focuses on GWP to facilitate a comparative evaluation of climate impact among energy carriers. The GREET model simulates emissions across the supply chain, quantifying them as CO_2-equivalent emissions [64]. The GWP values are determined over a 100-year period as specified in the Fifth Assessment Report (AR5) of the Intergovernmental Panel on Climate Change (IPCC) [65].

The concluding phase of the life cycle assessment (LCA), interpretation, entails assessing outcomes from life cycle inventor (LCI) and life cycle impact assessment

8.3 Life Cycle Assessment

(LCIA) to identify the primary sources of total emissions. This step identifies critical areas for reducing GHG emissions, informing decisions in energy production, transportation, and consumption. By synthesizing all these phases, the study delivers a thorough environmental assessment of various energy carriers, elucidating the most sustainable and efficient routes for energy transit and utilization.

The suggested supply chain for energy carriers includes standard activities such as feedstock generation, fuel production, storage, transportation, and consumption, as depicted in Fig. 8.14. The research accounts for emissions generated by (i) the combustion of process fuels that supply energy for operations, and (ii) leaks and boil-off gas (BOG) related to the production, loading, storage, transit, and consumption of energy carriers. The emissions for each method evaluated in this study are detailed in the supplementary file.

Empirical data are utilized to enhance the relevance and validity of the study. The suggested approach's restriction about whether LCA information sufficiently guides decision-makers includes elements such as infrastructure requirements, maintenance, fuel availability, and fuel cost. These elements are excluded from the life cycle assessment (LCA). Nevertheless, supplementary tools may encompass broader

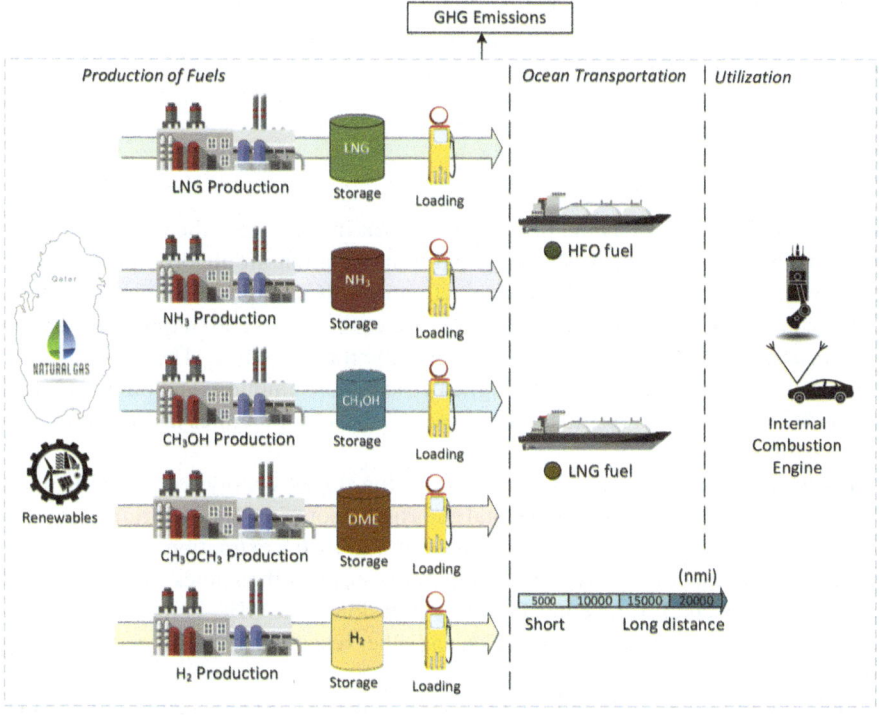

Fig. 8.14 Process flowchart showing various stages within the system boundary (from [59], licensed under CC-BY 4.0)

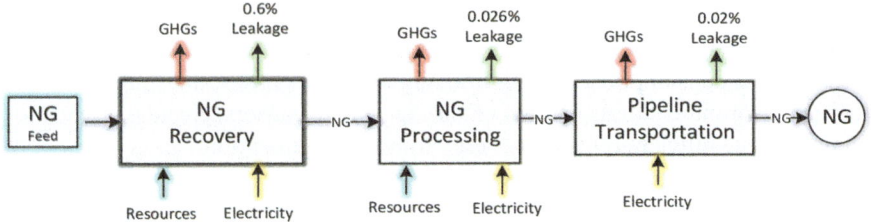

Fig. 8.15 Natural gas production process main inputs and outputs [59]

dimensions in a multicriteria decision framework, including life cycle costing and sustainability evaluation [66]. Nonetheless, LCA typically serves as the preliminary phase for such examination. Consequently, the LCA methodology can be augmented with additional instruments for thorough evaluations. The results of the LCA may not directly influence the final decision on the selection of an energy carrier, but they can serve as supporting arguments for that decision.

All data about energy requirements and leakage rates in processes are derived from GREET inventory data, whereas BOG generation rates are sourced from comprehensive prior publications [11, 12]. The subsequent subsections delineate the supply chains for the fuels, specifying input and output resources, electricity demands, leakage rates, and BOG rates for each process.

Production of Energy Carriers

Since natural gas serves as the primary feedstock in this study, the emissions associated with its extraction, processing, and transportation are incorporated into the total emissions calculation for the production of LNG, liquid ammonia, methanol, DME, and liquid hydrogen. Figure 8.15 illustrates the inputs and outputs across each subprocess, from natural gas recovery to pipeline transportation. The key inputs for these stages include natural gas flaring, electricity consumption, and other essential resources such as water, residual oil for boilers, diesel for engines and turbines, and gasoline blend-stock.

During the natural gas recovery and processing stages, energy consumption plays a significant role in determining the total emissions footprint. The electricity required to produce 1 GJ of natural gas is estimated at 270 kJ for extraction and 862 kJ for processing [44]. Additionally, due to internal system inefficiencies, some unrecovered gas leaks occur, further contributing to emissions. The GHG emissions resulting from energy consumption, flaring, leaks, and other resource usage are systematically accounted for in the LCA model.

Once natural gas is extracted and processed, it is transported via an 80 km pipeline to various industrial processing plants where it is converted into different liquefied energy carriers. The electricity required to pump natural gas over this distance also

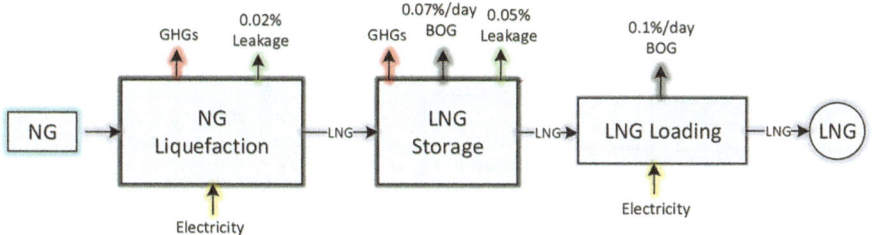

Fig. 8.16 Sources of emissions, leaks, and BOGs during LNG production [59]

generates emissions, which are included in the total life cycle emissions. These emissions, combined with those generated from the production processes of LNG, liquid ammonia, methanol, DME, and liquid hydrogen, provide a comprehensive assessment of the environmental impact associated with producing these energy carriers. By considering all upstream emissions from natural gas recovery to industrial conversion, this study ensures an accurate and holistic evaluation of the full supply chain emissions for each energy carrier.

LNG

The LNG supply chain commences with the liquefaction of natural gas, a procedure that entails cooling the gas to − 162 °C to transform it into a liquid form for enhanced storage and transport efficiency. The principal sources of greenhouse gas emissions during LNG production stem from fuel consumption necessary for the operation of liquefaction apparatus and gas losses resulting from venting and leaks in pipelines and equipment. These emissions are depicted as electrical inputs and gas losses in Fig. 8.16 [67].

Upon production, LNG is moved to on-site storage tanks specifically engineered to accommodate substantial volumes of LNG prior to being loaded aboard ocean tankers for exportation. Throughout the storage and loading periods, boil-off gas (BOG) is produced at a consistent rate, resulting in increased emissions. The GREET model incorporates integrated inventory data for these processes, facilitating a comprehensive evaluation of emissions produced at each phase.

The aggregate emissions across the LNG supply chain—from production to storage and loading—are calculated by aggregating the emissions resulting from fuel use, gas leaks, and boil-off gas generation. This study offers a thorough assessment of the environmental impact of LNG production and transportation by considering all emission sources during these phases. Comprehending these emission contributions is essential for recognizing potential mitigation solutions, including BOG capture and usage, enhancements in energy efficiency, and technology for leak reduction, to improve the sustainability of LNG as an energy carrier.

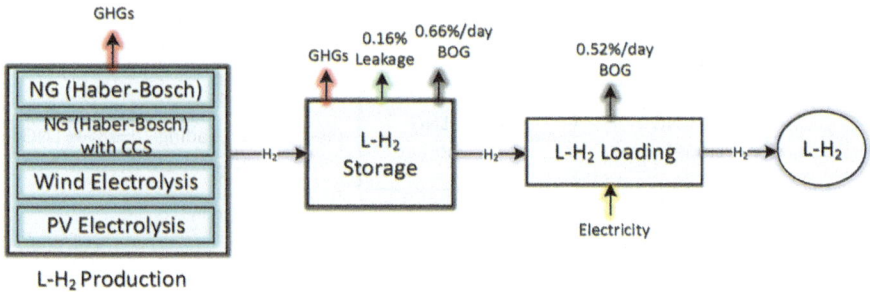

Fig. 8.17 Sources of emissions in L–H2 production [59]

Liquid Hydrogen

Hydrogen can be generated either via the reforming of natural gas or from renewable sources, including wind and solar energy, as depicted in Fig. 8.17. The hydrogen production from natural gas reforming adheres to a predefined model in GREET, which excludes considerations for CO_2 sequestration or carbon capture. This study utilizes GHG emissions from hydrogen synthesis through natural gas reforming, sourced directly from GREET inventory data. These emissions predominantly stem from fuel consumption necessary for operating turbines and motors, together with fuel leakages that transpire during processing.

Conversely, hydrogen can be generated via water electrolysis utilizing renewable energy sources like wind and solar (PV) energy. The greenhouse gas emissions from these renewable hydrogen production systems are sourced from prior research. Reported data indicates that the greenhouse gas emissions linked to liquid hydrogen (L–H_2) production using wind-powered electrolysis are 0.97 kg CO_2-eq per kg of hydrogen, while PV-powered electrolysis yields 2.4 kg CO_2-eq per kg of hydrogen [68].

A key component affecting hydrogen's environmental impact is its role as an indirect greenhouse gas. In contrast to direct greenhouse gases like CO_2 or CH_4, hydrogen emissions into the atmosphere can modify the composition of tropospheric free radicals, including methane (CH_4), hydroxyl radicals (OH), and molecular hydrogen (H_2). These alterations can augment global radiative forcing, hence exacerbating climate change. Consequently, although hydrogen is sometimes seen as a clean energy carrier, its emissions into the atmosphere during production, storage, and transportation require meticulous management. In the absence of efficient hydrogen capture and containment methods, its unregulated discharge may result in adverse environmental effects.

Upon production, liquid hydrogen (L–H_2) is conveyed to storage tanks prior to being loaded onto cargo tankers for transportation. During the storage and loading phases, a fraction of L–H_2 is lost as boil-off gas (BOG) and through leaks, hence exacerbating hydrogen losses and emissions. These losses must be alleviated through enhanced insulation, sophisticated reliquefication technology, and effective hydrogen containment measures. Integrating hydrogen leak prevention devices into the supply

8.3 Life Cycle Assessment

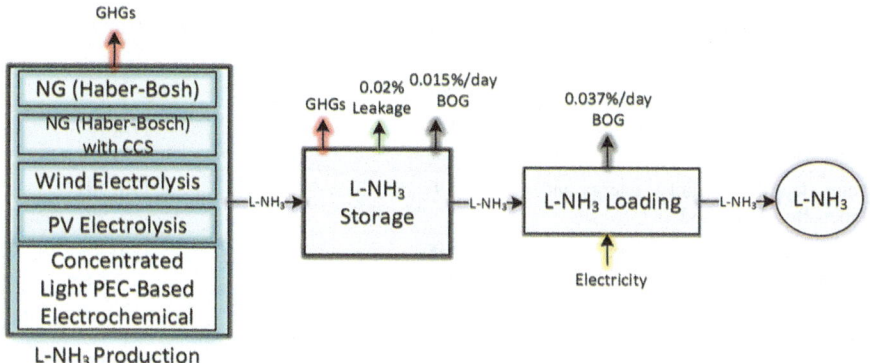

Fig. 8.18 Sources of emissions in ammonia production [59]

chain maximizes the environmental advantages of utilizing hydrogen as a clean fuel, facilitating a more sustainable shift from fossil fuels [69].

Liquid Ammonia

The liquid ammonia (L–NH_3) production pathway can be derived from natural gas reforming or renewable energy sources, as shown in Fig. 8.18. This study investigates GHG emissions from different L–NH_3 production methods to assess their environmental impact across the supply chain. The most widely used method for ammonia production is the Haber–Bosch process, which synthesizes ammonia by combining nitrogen from the air with hydrogen derived from either natural gas or water electrolysis. This method is responsible for producing the majority of ammonia worldwide and accounts for approximately 3–5% of total global natural gas consumption [70].

To enhance sustainability, integrating carbon capture and storage (CCS) facilities into ammonia production plants has been explored both in literature and at an industrial scale. This study evaluates GHG emissions from L–NH_3 production both with and without CCS technology. It has been found that incorporating CCS into conventional natural gas-based ammonia production can lead to a 25% reduction in GHG emissions [71]. This reduction percentage is applied in this study to estimate total emissions from L–NH_3 production when CCS is implemented. The GREET database provides inventory data for estimating emissions from ammonia production using natural gas reforming. However, ammonia production through water electrolysis (using PV or wind energy) and photoelectrochemical (PEC) methods are not available in GREET; thus, data from previous studies is utilized.

Table 8.11 presents a summary of GHG emissions for ammonia production from renewable sources. When using wind energy for water electrolysis combined with the Haber–Bosch process, GHG emissions are 0.4 kg CO_2-eq per kg of NH_3. In contrast, water electrolysis via PV energy results in higher emissions at 0.9 kg CO_2-eq per kg of NH_3, while concentrated light PEC-based electrochemical methods produce 1.06

Table 8.11 Summary of GHG emissions of renewable-based ammonia production methods

Method	GHG emissions (kg of CO_2 eq/ kg of NH_3)	References
Water electrolysis via wind energy and Haber–Bosch method	0.4	[72, 73]
Water electrolysis via PV energy and Haber–Bosch method	0.9	[73, 74]
Concentrated light PEC-based electrochemical	1.06	[75]

kg CO_2-eq per kg of NH_3. These emissions are then integrated into the LCA model to estimate total emissions from the entire supply chain of L–NH_3.

After production, L–NH_3 is transferred to storage tanks before being loaded onto shipping vessels for transport. One important factor influencing GHG emissions in ammonia storage is the difference in temperature between storage conditions and the surrounding environment. Since L–NH_3 is stored at − 34 °C, whereas LNG requires storage at − 160 °C, ammonia experiences significantly lower heat transfer into storage tanks. As a result, the leakage rate and BOG rate of L–NH_3 storage are 2.5 and 5 times lower than LNG storage, respectively. The emissions caused by ammonia losses during storage, as well as fuel requirements for production and loading operations, are factored into the total GHG emission estimates.

By analyzing various production pathways, transportation methods, and storage losses, this study provides a comprehensive assessment of the environmental footprint of ammonia as an energy carrier. The findings highlight the potential of renewable-based ammonia production to reduce GHG emissions, while also emphasizing the importance of minimizing storage losses and optimizing energy efficiency across the supply chain.

Methanol and DME

The GREET software is employed to calculate and evaluate GHG emissions during various phases of methanol and DME production, as depicted in Fig. 8.19. This case study assesses greenhouse gas emissions from two main production methods: natural gas reforming (via steam methane reforming—SMR) and biomass gasification. The GREET database offers emissions inventory data for both processes, facilitating a comparative evaluation of their environmental impact.

The methanol production process commences with the generation of syngas, produced by a steam reformer. The generated syngas is subjected to compression, with pressures varying from 30 to 100 bar. The compressed gas is subsequently directed through a catalytic converter at a reaction temperature of around 300 °C. The gaseous byproduct from this process is channeled into a condenser, where thermal recovery produces steam. The final step entails the separation of methanol from the condensed product in a knockout pot, thereby eliminating unreacted gases and isolating pure methanol for storage and subsequent processing [44].

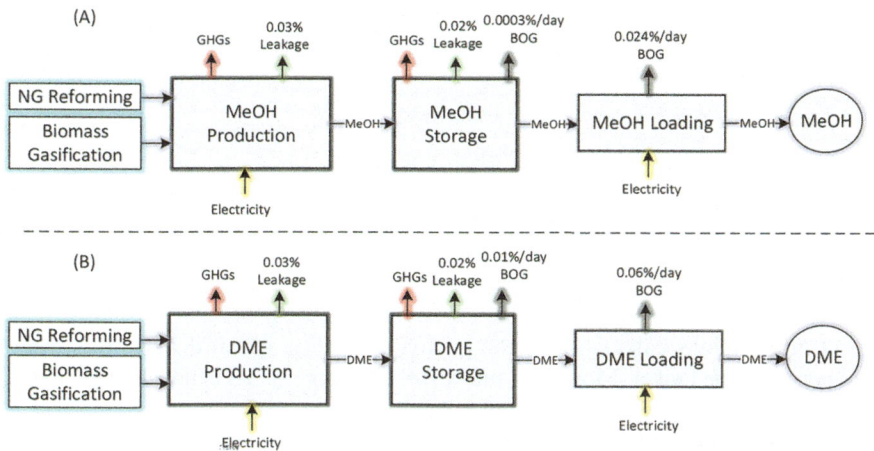

Fig. 8.19 Sources of emissions in **a** methanol and **b** DME production pathways [59]

DME manufacture may utilize either a single-stage or a two-stage synthesis procedure. The single-stage process directly transforms syngas into DME via methanol dehydration, whereas the two-stage process initially generates methanol, which is subsequently transformed into DME. This study employs a two-stage process to quantify greenhouse gas emissions across the DME life cycle. This method offers a comprehensive analysis of emissions produced at each phase, including syngas generation, compression, methanol synthesis, and the ultimate DME conversion [76].

This study compares natural gas-based and biomass-based production methods, emphasizing the disparities in greenhouse gas emissions between fossil fuel-derived methanol and DME and their renewable alternatives. These insights facilitate the comprehension of environmental trade-offs and the identification of more sustainable production alternatives for methanol and DME as low-carbon energy carriers.

Transportation and Utilization of Energy Carriers

Greenhouse gas (GHG) emissions are assessed across the whole fuel supply chain in Qatar, encompassing raw material extraction, processing, production, storage, and loading onto maritime tankers. After fuels are generated and stored on-site, the subsequent essential phase in the supply chain is maritime transportation, during which substantial volumes of fuels are conveyed over extensive distances utilizing specialized shipping tankers. In 2019, Qatar delivered 77.9 million tons of natural gas in LNG form via ocean tankers, underscoring the importance of maritime transit in global energy trade [77].

This study models the transportation of fuels by ocean tankers to assess the related greenhouse gas emissions. Table 8.12 delineates essential specifications for ocean tankers, outlining the criteria for fuel transportation across varying distances. A key

Table 8.12 Ocean tanker specific parameters

Ocean tanker specific parameters	Value
Payload (kg)	60,000,000
Average speed (m/s)	8.94
Load factor to	0.83
Load factor from	0.70
Brake specific fuel consumption (g/kWh)	195
Typical horsepower (hp)	9070

aspect in emissions calculation is the payload capacity, which denotes the overall load borne by the tanker. Moreover, the vessel's average speed is taken into account, as it influences both trip time and fuel efficiency.

The study includes two load variables to account for energy intensity and fuel usage in transportation. "Load factor to" denotes the percentage of engine power employed during the transit from the point of origin to the destination. "Load factor from" denotes the percentage of engine power utilized during the return journey from the destination to the original site.

Additional essential parameters encompass brake-specific fuel consumption (BSFC) and standard horsepower ratings, which ascertain the overall energy necessary for propulsion. The GREET software incorporates these factors and features a built-in method for calculating energy intensity and emissions produced by the ocean tanker.

The greenhouse gas emissions produced by the fuel utilized to operate the tanker significantly contribute to total transportation-related emissions. Emissions are contingent upon variables including vessel size, fuel type, engine efficiency, and travel distance. The study offers a comprehensive evaluation of emissions from water transportation by integrating these aspects, facilitating the identification of options to mitigate maritime emissions via alternative fuels, enhanced energy efficiency, or optimal logistics in fuel transfer.

Additionally, BOG generation rates for various fuels in ship tanks are documented in our prior work [12]. Table 8.13 presents BOG rates during the maritime shipping phase. The BOG rates are time-dependent loss rates, indicating that the duration of journey is contingent upon the distance and the average speed of the ocean tanker. The emissions resulting from BOG losses are included in the overall GHG emissions for each supply chain.

The overall GHG emissions during the transportation phase are derived from the aggregation of fuel requirements for each ocean tanker plus mass losses due to BOG and leaks. A sensitivity analysis is conducted to assess overall GHG emissions resulting from fuel production and transportation over varying distances by different types of fueled ocean tankers. Diesel and heavy fuel oil are typically utilized as fuel for shipping tankers. LNG is increasingly utilized as fuel for vessels to mitigate emissions. A comparison analysis is provided between heavy fuel oil (HFO) and LNG utilized as fuel for ocean ships [78].

Table 8.13 Daily BOG rates in ocean tankers during ocean transportation for various energy carriers

Energy carrier	BOG/day (%)
LNG	0.12
L–NH$_3$	0.025
MeOH	0.005
DME	0.017
L–H$_2$	1.06

Source [12]

Upon transportation of the liquefied energy carriers to the designated area, they are deemed suitable for use in internal combustion engine vehicles (ICEVs) for propulsion. The combustion and emission parameters of LNG, DME, methanol, and L–H$_2$ are derived from GREET software, which includes fueled internal combustion engines for each type, excluding L–NH$_3$. The combustion of L–NH$_3$ is determined by its chemical reaction and L–H$_2$ data. The emissions produced during the use phase are subsequently combined with emissions from manufacturing and transportation to calculate the overall greenhouse gas emissions of the entire supply chain for these fuels.

8.3.5.3 Results

The GREET inventory data encompasses greenhouse gas (GHG) emissions from multiple fuel generation paths. The fuel pathways in this study were adjusted to conform to particular parameters, encompassing feedstock sources, electricity sources, travel distances, and boil-off gas (BOG) rates. Following the implementation of these modifications, greenhouse gas emissions were assessed throughout various stages of the fuel supply chain.

A validation study was done to confirm the accuracy and reliability of the predicted emissions by comparing the results with emissions data from prior studies. Table 8.14 delineates a comparative analysis of the literature, concentrating on the greenhouse gas emissions produced throughout the production phase of energy carriers originating from natural gas.

The study estimates GHG emissions for LNG production at 10.29 g/MJ, aligning with the literature range of 9–13.3 g/MJ. Correspondingly, emissions from methanol production are estimated at 20.10 g/MJ, which closely corresponds with previously documented values (20–24 g/MJ). The results for DME production (21.11 g/MJ) align with the extensive range documented in literature (12.2–47.2 g/MJ), suggesting discrepancies attributable to variances in feedstock sources and production efficiency.

The generation of liquid hydrogen results in greenhouse gas emissions of 119.92 g/MJ, closely aligning with reference values of 120.4 g/MJ, so affirming the significant energy intensity associated with hydrogen synthesis using steam methane reforming (SMR) in the absence of carbon capture. The projected emissions for liquid ammonia

Table 8.14 Literature comparison of GHG emissions of energy carriers produced from natural gas in their production phases [59]

	GHG emissions (this study)	Reference 1	Reference 2	Reference 3
Production phase				
LNG (g/MJ)	10.29	9 [79]	10.70 [80]	13.3 [81]
Methanol (g/MJ)	20.10	21.97 [82]	24.0 [83]	20 [84]
DME (g/MJ)	21.11	12.2–36.7 [85]	16.3–47.2 [85]	
Liquid hydrogen (g/MJ)	119.92	120.4 [86]		
Liquid ammonia (g/MJ)	41.66	29.7 [87]	48 [88]	23.25–33.48 [89]
Utilization phase				
LNG (g/MJ)	57.5	56 [90]	60 [91]	
Methanol (g/MJ)	68.64	69 [91]	68 [92]	68 [93]
DME (g/MJ)	67.81	66 [92]	67 [93]	67 [94]
Liquid hydrogen (g/MJ)	0.54	0 [91]	0 [95]	0 [94]
Liquid ammonia (g/MJ)	0.54	0 [96]	0 [95]	

production are 41.66 g/MJ, consistent with reported values between 29.7 and 48 g/MJ.

In the utilization phase, where energy carriers are utilized in internal combustion engines (ICEs), the projected emissions for LNG (57.5 g/MJ) align well with previously published values (56–60 g/MJ). The emissions from methanol and DME usage are estimated at 68.64 g/MJ and 67.81 g/MJ, respectively, aligning with values documented in several research.

The analysis verifies that the use of liquid hydrogen and liquid ammonia results in nearly nil direct CO_2 emissions, consistent with prior literature findings. The 0.54 g/MJ measurement indicates negligible indirect emissions, perhaps due to trace fuel contaminants or fluctuations in combustion efficiency.

The findings demonstrate that the projected GHG emissions correspond closely with the current literature, validating the reliability and precision of the study's LCA approach and the adjustments implemented in GREET routes. The discrepancies noted in certain fuel pathways, especially for DME and ammonia, underscore the impact of production techniques, process efficiency, and feedstock variances among various studies.

This case study validates production and utilization emissions, ensuring that the LCA model appropriately represents real-world situations, so establishing it as a dependable instrument for evaluating the environmental impact of different energy carriers. This comparison method offers significant insights into options for emission reduction, including the incorporation of carbon capture technology in hydrogen and

8.3 Life Cycle Assessment

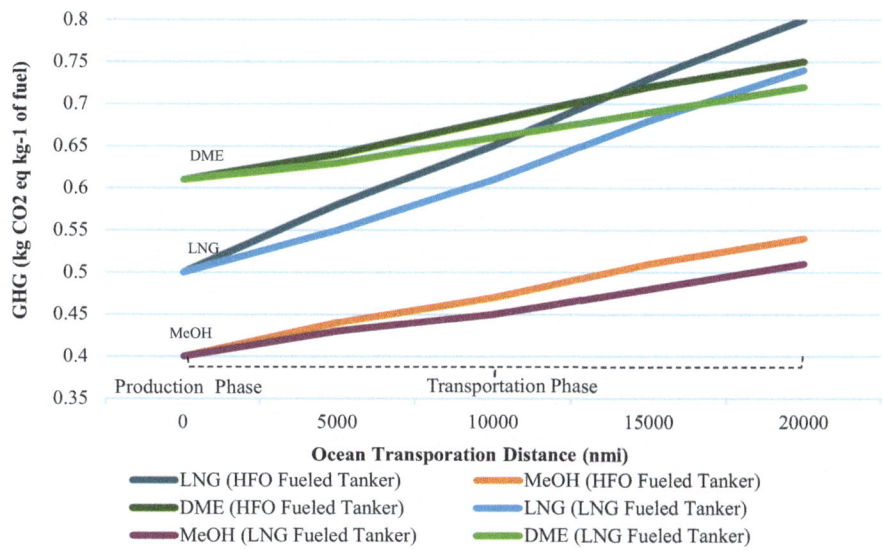

Fig. 8.20 GHG emissions generated from production and transportation of LNG, DME, and MeOH by two ocean tankers: (–) HFO-fueled tanker and (– – –) LNG-fueled tanker [59]

ammonia generation, as well as the optimization of process efficiency in methanol and DME synthesis.

Figure 8.20 presents a comparative comparison of greenhouse gas emissions from LNG, DME, and methanol during their manufacturing and transportation phases. Of the three fuels, methanol exhibits the lowest greenhouse gas emissions during manufacture. The production of 1 kg of methanol from natural gas entails steam reforming in conjunction with partial oxidation, yielding roughly 0.4 kg of CO_2-equivalent emissions. This emission level is 25% lower than that of LNG and 50% lower than DME, establishing methanol as the least carbon-intensive alternative during the production phase.

Following manufacture, these fuels are generally delivered internationally by ocean tankers, a practice that additionally contributes to overall emissions. The transportation phase produces emissions chiefly from two sources: the gasoline utilized by the ocean tanker and the boil-off gas (BOG) emitted during transit. The fuel type utilized by the tanker substantially influences emissions. An HFO-fueled ocean tanker generates greater emissions than an LNG-fueled tanker. The overall well-to-propeller global warming potential of HFO exceeds that of LNG by 23% when utilized as a shipping fuel for equivalent transit distances [97].

A further aspect affecting GHG emissions in ocean transport is the formation of BOG. The daily boil-off gas (BOG) generation rate of LNG is 0.12%, indicating that increased transportation distance correlates with higher emissions from LNG. LNG is less environmentally advantageous for long-distance transportation as compared to methanol and DME. While DME initially exhibits greater emissions than LNG, transportation distances above 14,000 nautical miles (nmi) result in LNG emissions exceeding those of DME, because to cumulative BOG losses and heightened fuel usage.

Considering solely the manufacturing and transportation stages, methanol and DME exhibit greater environmental advantages than LNG. Methanol exhibits the lowest emissions during production and does not incur BOG losses during maritime transit, positioning it as a viable option for minimizing emissions in long-distance energy transfer. Nonetheless, supplementary emissions arise during the utilization phase, which are addressed in the subsequent sections. This comparison underscores the necessity for cleaner transportation fuels and the prospective advantages of alternative low-carbon energy carriers like methanol and DME in global fuel supply networks.

To adhere to environmental rules, it is essential to evaluate the entire supply chain of fuels, including manufacturing, transportation, and utilization stages. Figure 8.21 depicts the greenhouse gas emissions for the complete supply chain of LNG, DME, and methanol, quantified per kilogram (kg) and per megajoule (MJ) of energy. The use phase represents the predominant portion of GHG emissions for each fuel type. In the LNG supply chain, the use phase constitutes about 77% of total emissions when transporting LNG across a distance of 20,000 nautical miles (nmi). The production phase accounts for around 15% of total emissions, with the remainder originating from the transportation phase.

Examining the entire fuel supply chain reveals significant disparities in greenhouse gas emissions among the three energy carriers. When methanol and DME are derived from natural gas, delivered over 5000 nautical miles via an HFO-fueled ocean tanker, and employed in an internal combustion engine, they realize a reduction in GHG emissions of 45% and 20%, respectively, in comparison to LNG. This suggests that methanol and DME may be more environmentally advantageous for minimizing GHG emissions over shorter transportation distances.

Nonetheless, discrepancies in emissions exist when evaluated by energy content (per MJ). Despite LNG producing fewer emissions owing to its reduced carbon content per energy unit, it generates around 70 g of CO_2-equivalent per MJ along its entire supply chain. Methanol emits 90 g of CO_2-equivalent per MJ, whereas DME emits 89 g of CO_2-equivalent per MJ. The elevated emissions of methanol per energy unit result from its worse heating value and increased density relative to LNG and DME.

In summary, when evaluating overall greenhouse gas emissions across the whole supply chain, LNG produces the lowest emissions per energy unit. Nevertheless, when assessing emissions per unit mass, methanol has the lowest emissions of the three fuels. This signifies that the ideal fuel selection is contingent upon the application and priorities—whether reducing emissions per unit of energy or per unit of mass is paramount. These insights underscore the necessity for a comprehensive strategy in fuel supply chain management to attain the most environmentally sustainable results.

Liquid hydrogen (L–H_2) and liquid ammonia (L–NH_3) are two of the most promising carbon–neutral energy carriers for transport and usage. Qatar's extensive natural gas reserves enable the production of energy carriers via various methods,

8.3 Life Cycle Assessment

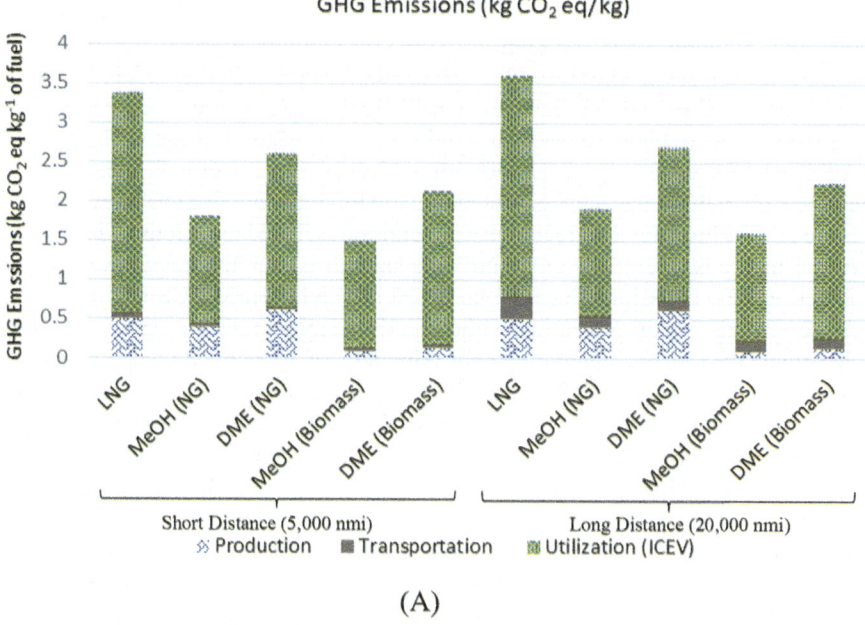

(A)

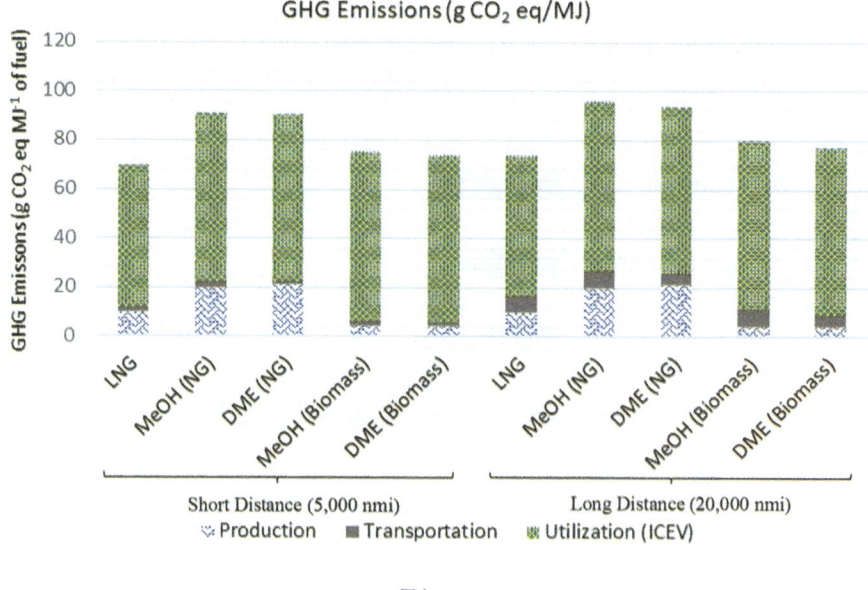

(B)

Fig. 8.21 Total GHG emissions from production, transportation, and utilization of LNG, DME, and methanol for short distances and long distances, **a** per unit mass (kg CO_2 eq kg^{-1} of fuel) and **b** per unit energy (gCO_2 eq MJ^{-1} of fuel) (from [59], licensed under CC-BY 4.0)

including steam methane reforming (SMR) with or without carbon capture and storage (CCS), natural gas cracking utilizing concentrated solar energy, or water/seawater electrolysis powered by renewable energy sources like solar photovoltaic or wind energy. This study evaluates renewable hydrogen generation via water electrolysis powered by photovoltaic and wind energy, resulting in gaseous hydrogen. Nonetheless, the electricity necessary for hydrogen liquefaction is presumed to originate from natural gas, hence influencing the total greenhouse gas emissions.

Figure 8.22 illustrates the greenhouse gas emissions linked to various L–H_2 manufacturing methods, including those utilizing natural gas as the energy source for liquefaction. The emissions fluctuate based on the inclusion of CCS technology in steam reforming. Steam reforming without carbon capture and storage is the most carbon-intensive process, generating approximately 14.39 kg of CO_2-equivalent per kilogram of low-hydrogen (L–H_2). Nonetheless, the implementation of CCS in steam reforming markedly decreases emissions, resulting in a 59% reduction in total greenhouse gas emissions. Conversely, wind-powered hydrogen production is the most carbon-efficient method, emitting merely 3.6 kg of CO_2-equivalent per kilogram of L–H_2.

The emissions disparity between renewable hydrogen production using wind-based electrolysis and natural gas reforming technologies is significant. Hydrogen produced from wind energy generates 40% less CO_2 compared to steam methane reforming (SMR) with carbon capture and storage (CCS) and 75% less than SMR without CCS. This underscores the environmental advantages of hydrogen production from renewable sources; yet, its extensive implementation is contingent upon factors including cost, infrastructural accessibility, and energy efficiency limitations.

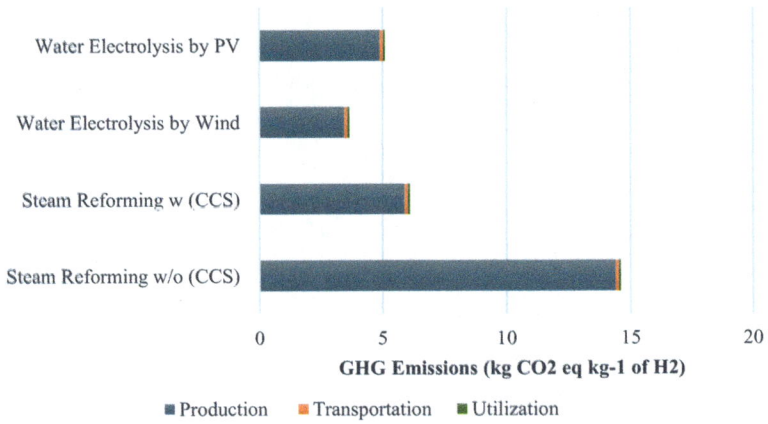

Fig. 8.22 Total GHG emissions from different L–H_2 production methods covering emissions in production, 20,000 nmi transportation by HFO-fueled ocean tanker, and utilization in internal combustion engine processes

8.3 Life Cycle Assessment

These findings underscore the essential function of CCS technology in mitigating GHG emissions from hydrogen generation derived from natural gas. Although renewable hydrogen production using solar or wind electrolysis is the most environmentally benign method, the practicality of large-scale deployment is a significant hurdle. Consequently, when natural gas is the principal feedstock, the use of CCS technology is imperative for attaining significant GHG reductions and conforming to global climate objectives. By utilizing natural gas with carbon capture and storage alongside renewable energy, Qatar can develop a low-carbon hydrogen economy, facilitating future clean energy markets while reducing environmental effect.

In the analysis of the L–H_2 supply chain, it is evident that the predominant source of GHG emissions arises during the production phase, specifically from the reforming and liquefaction operations. The production of 1 kg of gaseous hydrogen necessitates around 50 kWh of power, along with an extra 10 kWh for liquefaction [98]. This underscores the essential influence of power generation sources on the environmental ramifications of hydrogen production. In nations with substantial natural gas reserves, like Qatar, power is predominantly produced from natural gas, which considerably impacts the carbon footprint of L–H_2 generation.

Figure 8.23 depicts greenhouse gas emissions from different low-carbon hydrogen generation technologies that utilize natural gas-derived energy for both reforming and liquefaction operations. Furthermore, it delineates the emissions associated with L–H_2 production systems that employ renewable electricity (solar PV) just for liquefaction. The findings demonstrate that transitioning from natural gas-derived electricity to solar photovoltaic systems decreases greenhouse gas emissions by roughly 2.3 kg CO_2-equivalent per kilogram of L–H_2 in hydrogen manufacturing processes.

In Qatar, where L–H_2 is generated through steam reforming with carbon capture and storage (CCS) with natural gas-derived electricity, and subsequently transported

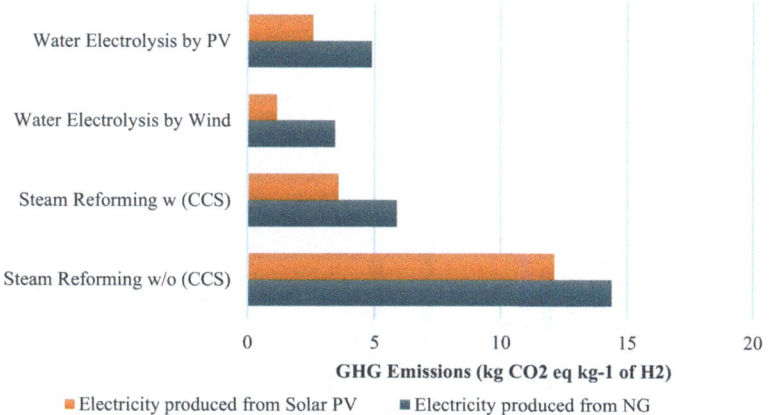

Fig. 8.23 Comparison of produced GHG emissions from the L–H_2 supply chain (production, storage, and 20,000 nmi ocean transportation) that uses generated electricity either by natural gas or by solar PV

via an HFO-powered ocean tanker over a distance of 20,000 nautical miles, the total emissions reach 5.89 kg CO_2-equivalent per kg of L–H_2. Conversely, employing solar PV-generated electricity in the liquefaction process for the identical supply chain scenario leads to a 39% decrease in overall GHG emissions.

These findings highlight the significance of utilizing renewable electricity in hydrogen generation to attain considerable decreases in greenhouse gas emissions. Although natural gas reforming with carbon capture and storage is a feasible low-carbon alternative, shifting to renewable energy sources for power generation during the production phase can substantially enhance the environmental impact of hydrogen as a clean energy carrier. Consequently, including solar or wind energy into hydrogen supply chains presents a viable approach for nations seeking to decarbonize hydrogen production and conform to global sustainability objectives.

Liquid ammonia (L–NH_3) can be synthesized using natural gas reforming or processes utilizing renewable energy. Figure 8.24 depicts the greenhouse gas emissions linked to different L–NH_3 production methods, along with emissions from maritime transport spanning distances of 0–20,000 nautical miles (nmi). The analysis indicates that a 5000 nmi increase in transport distance results in only a 1% spike in overall GHG emissions, predominantly attributed to the fuel consumption of the HFO-powered shipping tanker.

In Qatar, the production of L–NH_3 from natural gas reforming without carbon capture and storage (CCS) generates emissions of roughly 2.24 kg CO_2-equivalent per kg of L–NH_3. Integrating CCS technology decreases GHG emissions by 24%, rendering it a more environmentally advantageous choice. Conversely, the synthesis

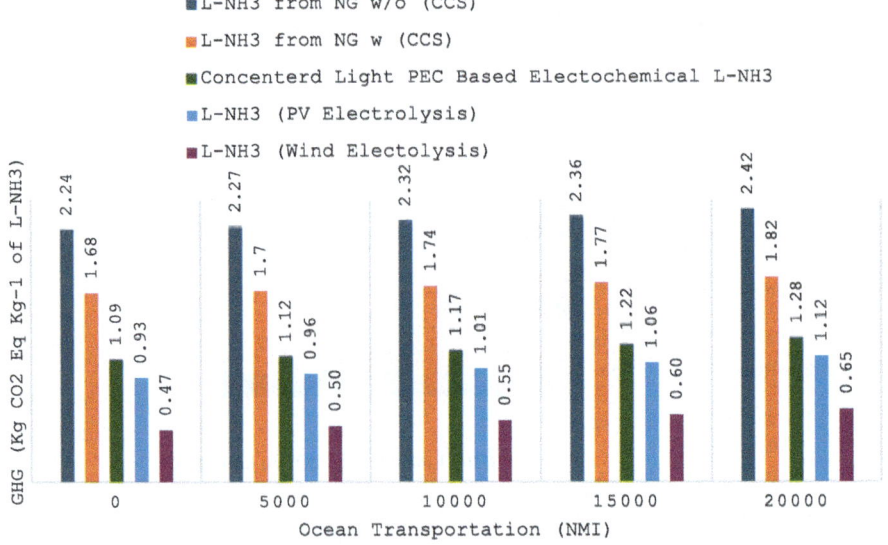

Fig. 8.24 GHG emissions generated from the production of L–NH_3 by different production methods and transportation over various distances by HFO-fueled ocean tanker

8.3 Life Cycle Assessment

of L–NH$_3$ through the electrolysis of water with renewable energy sources results in markedly reduced emissions—0.47 kg CO$_2$-eq per kg of L–NH$_3$ when employing wind energy and 0.93 kg CO$_2$-eq per kg of L–NH$_3$ when utilizing solar photovoltaic technology.

The concentrated light photoelectrochemical (PEC) process is a novel manufacturing technology that is attracting interest for its ability to generate ammonia with reduced carbon emissions. The greenhouse gas emissions for the production of 1 kg of L–NH$_3$ with PEC technology are predicted to be 1.09 kg CO$_2$-equivalent [99]. PEC-based ammonia production results in 35% lower CO$_2$ emissions than natural gas reforming with CCS, although 15% higher emissions than water electrolysis with solar PV.

These findings underscore the significance of manufacturing technique selection in ascertaining the total carbon footprint of ammonia. Although natural gas reforming with carbon capture and storage can markedly diminish emissions relative to conventional reforming, renewable ammonia production techniques—such as electrolysis and photoelectrochemical technology—present minimal emissions and are consistent with long-term decarbonization objectives. Furthermore, while transportation emissions constitute a negligible portion of overall L–NH$_3$ supply chain emissions, the most significant potential for emission reductions exists in the production phase. By transitioning to renewable-based ammonia production, nations such as Qatar can improve their sustainability initiatives and establish L–NH$_3$ as a feasible carbon-free energy carrier for international markets.

Considering the complete life cycle is crucial for attaining decarbonization, as indicated by the IMO [100]. Consequently, greenhouse gas emissions generated from L-NH$_3$ must be assessed along the entire supply chain. Figure 8.25 illustrate greenhouse gas emissions in various processes of the L–NH$_3$ supply chain per unit mass (kg) and per unit energy (MJ), respectively. Emissions predominantly occur during the production phase, regardless of the production methods utilized. Greenhouse gas emissions from the production phase constitute 90%, while transportation of L–NH$_3$ via HFO ocean tanker over about 20,000 nautical miles accounts for 8%, and its application in internal combustion engines represents 2% of the overall L–NH$_3$ emissions. Integrating a CCS facility into the standard ammonia production process (utilizing natural gas) results in an emission of around 100 g of CO$_2$ equivalent per MJ of L–NH$_3$, representing a 23% reduction compared to a facility lacking a CCS system. Conversely, the supply chain employing electrolyzers for L–NH$_3$ synthesis significantly lowers greenhouse gas emissions. Ammonia generation using wind electrolysis results in the minimal emissions of around 35.48 g of CO$_2$ equivalent per megajoule of L–NH$_3$.

Additionally, various emissions are generated during the production, transportation, and utilization of these fuels. NO$_x$ and SO$_x$ are detrimental emissions, and regulatory bodies are monitoring their repercussions. The emissions are derived from GREET inventory data, which may be monitored in the accompanying supplementary file. Table 8.15 delineates the NO$_x$ and SO$_x$ emissions created during the supply chain phases of various fuels derived from natural gas. The complete methanol supply chain exhibits the lowest NO$_x$ emissions, around 76.35 mg MJ^{-1}. Methanol presents

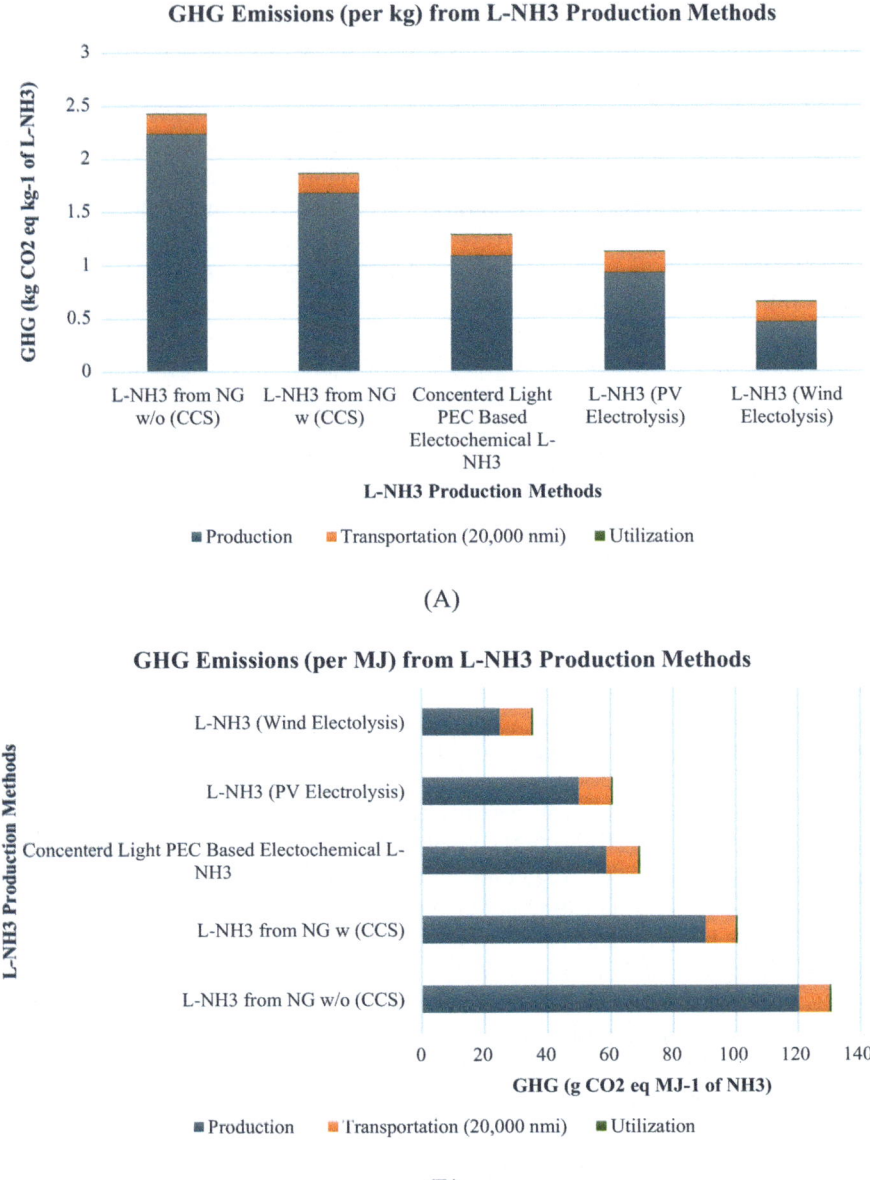

Fig. 8.25 Estimation of GHG emissions from different L–NH$_3$ production methods covering emissions in production, 20,000 nmi transportation by HFO-fueled ocean tanker, and utilization in an internal combustion engine, **a** per unit mass (kg), **b** per unit energy (MJ)

a viable alternative for adhering to environmental standards designed to mitigate dangerous NO_x emissions.

Figure 8.26 illustrates a comparative comparison of greenhouse gas emissions for methanol, DME, liquid hydrogen (L–H_2), and liquid ammonia (L–NH_3) derived from renewable sources or natural gas at various distances. The straight lines denote fuels generated from renewable energy sources. The findings demonstrate a decrease in greenhouse gas emissions relative to fuel sourced from natural gas. The whole life cycle of L–H_2 emits around 41.29 g of CO_2 per MJ of L–H_2, making it the cleanest fuel available. The total greenhouse gas emissions for the complete life cycle of methanol, DME, and L–NH_3 production from natural gas are 88.7, 88.9, and 90.9 g of CO_2 per MJ of fuel, respectively. The emissions are essentially identical, with ammonia exhibiting a somewhat higher value. Conversely, the renewable generation of these fuels indicates that overall GHG emissions for L–NH_3 amount to approximately 50.5 g of CO_2 per MJ of L–NH_3. This value is inferior to that of methanol and DME.

To determine the energy carrier with the lowest greenhouse gas emissions in Qatar, a comparison is conducted as illustrated in Fig. 8.27, measured in grams of CO_2 equivalent per kilogram and grams of CO_2 equivalent per megajoule, respectively. In the comparison of these fuels, the greenhouse gas emissions produced during the utilization phase constitute the majority of emissions in the LNG, DME, and MeOH supply chains, whereas the utilization emissions in the L–H_2 and L–NH_3 supply chains represent the least emissions. Despite LNG experiencing more energy loss due to elevated BOG rates, the greenhouse gas emissions from the entire LNG supply chain are lower than those from methanol and dimethyl ether produced through natural gas reforming, which exhibit reduced BOG rates. This pertains to reduced emissions during the production phase of LNG. Nonetheless, the production of DME and methanol via biomass gasification exhibits reduced greenhouse gas emissions during the production phase in comparison to the LNG production phase, rendering renewably produced DME and methanol more environmentally sustainable. Furthermore, the implementation of the CCS facility during the L–H_2 and L–NH_3 production phases decreases GHG emissions by 35% and 25%, respectively, hence contributing to a reduction in overall supply chain GHG emissions. The

Table 8.15 NO_x and SO_x emissions in the production, transportation (by HFO tanker for 20,000 nmi distance), and utilization (in internal combustion engine) phases of various fuels produced from natural gas

Fuel type	L–NH_3 (NG)		LNG		MeOH (NG)		DME (NG)		L–H_2 (NG)	
Unit	mg/MJ		mg/MJ		mg/MJ		mg/MJ		mg/MJ	
Emission types	NO_x	SO_x	NO_x	SO_x	NO_x	SO_x	NO_x	SO_x	NO_x	SO_x
Production	90.03	28.5	22.17	12.04	30.65	22.19	31.04	23.66	56.06	28.5
Transportation	34.9	15.5	35.1	21.04	37.1	20.24	41.24	25.99	34.9	15.5
Utilization	31.83	0.0	25.13	0.0	8.6	0.0	34.75	0.0	31.83	0.0

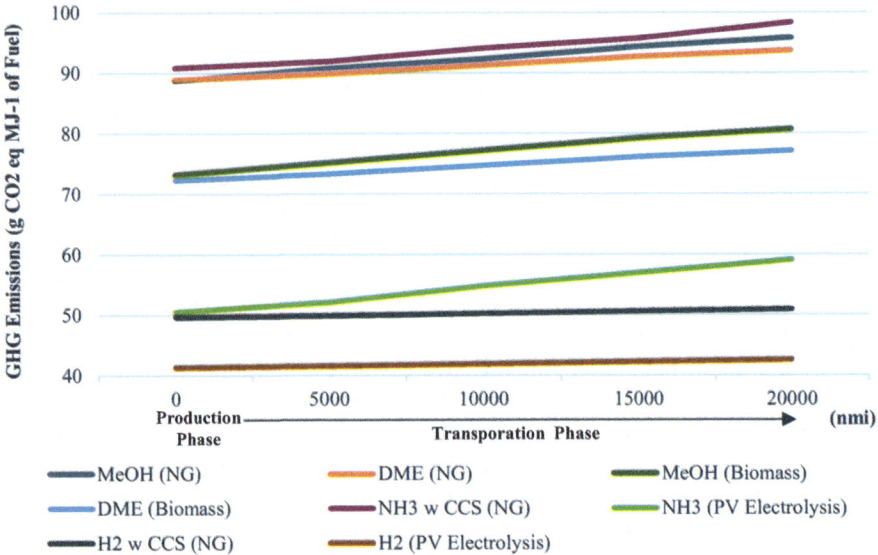

Fig. 8.26 Sensitivity analysis on various distances of a full life cycle of energy carriers derived from either renewables (solar PV and biomass) or natural gas covering production, transportation, and utilization phases

production of L–NH3 through natural gas reforming generates the highest greenhouse gas emissions; however, when ammonia is produced using renewable sources (photovoltaics), the overall greenhouse gas emissions across the whole supply chain are reduced compared to those of LNG, methanol, and dimethyl ether. Furthermore, the supply chain including manufacturing, 20,000 nautical miles of transportation via HFO-fueled ocean tanker, and use in an internal combustion engine of L–H$_2$ generated from solar photovoltaic sources exhibits the lowest greenhouse gas emissions among these fuels, around 42.5 g of CO_2 equivalent per megajoule of L–H$_2$.

8.3.5.4 Conclusions

Examining greenhouse gas emissions throughout the whole life cycle of energy carriers, encompassing production, transportation, and utilization, yields a thorough environmental evaluation by considering emissions from fuel leaks and boil-off gas generation. This study analyzes greenhouse gas emissions from LNG, DME, methanol, liquid hydrogen, and liquid ammonia, derived from either natural gas or renewable sources in Qatar. The results provide significant insights into the environmental effects of these energy carriers and the possibilities for emission reduction via alternative manufacturing methods and cleaner energy sources.

8.3 Life Cycle Assessment

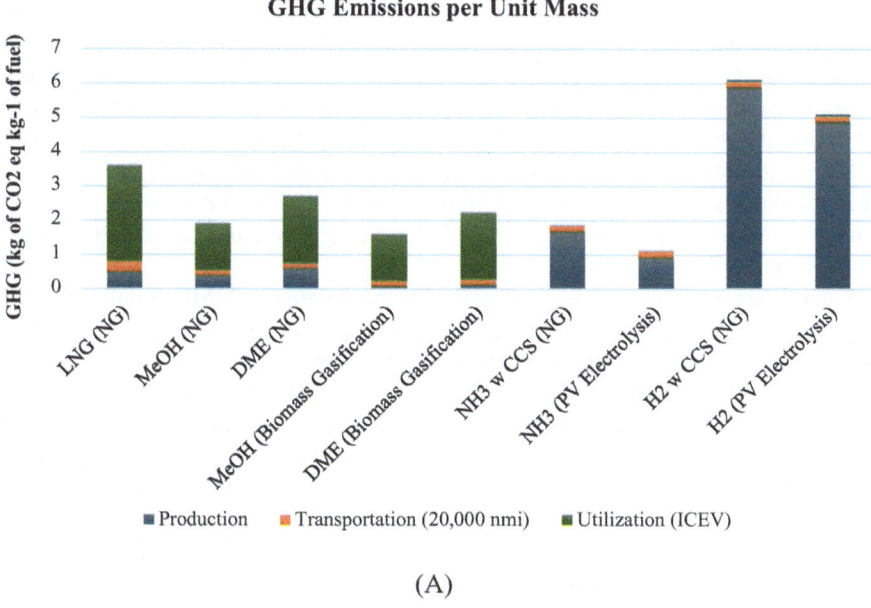

(A)

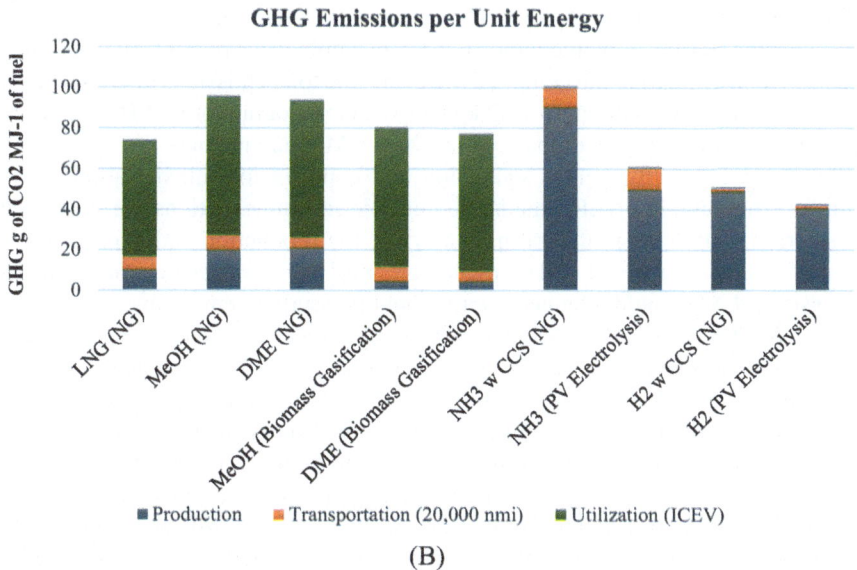

(B)

Fig. 8.27 Comparison of GHG emissions of different fuel supply chains, **a** per unit mass (kg), **b** per unit energy (MJ)

Comparing LNG, DME, and methanol reveals that methanol and DME exhibit reduced environmental consequences relative to LNG, focusing just on the manufacturing and transportation phases. The usage phase significantly impacts overall emissions, rendering comprehensive life cycle analysis essential. Methanol generated from natural gas reforming emits 95.8 g CO_2-eq/MJ, whereas DME emits 93.7 g CO_2-eq/MJ, representing increases of 30% and 27% over LNG, respectively. Nonetheless, when methanol and DME are derived from renewable sources (biomass), their greenhouse gas emissions diminish by 75 and 80% during production, establishing them as viable alternatives to LNG.

Regarding greenhouse gas emissions during transportation, DME initially exhibits higher emissions than LNG in the production phase; however, after maritime transportation distances surpass 14,000 nautical miles, LNG emerges as the more emissions-intensive fuel due to increased cumulative boil-off gas losses. The reduction of GHG emissions in L–H_2 production is significantly influenced by the source of power. Electricity generated from natural gas by steam methane reforming (SMR) with carbon capture and storage (CCS) produces 5.89 kg CO_2-equivalent every kilogram of low-carbon hydrogen (L–H_2). Replacing power generated from natural gas with solar photovoltaic systems decreases emissions by 40%, positioning renewable hydrogen as a viable low-carbon alternative.

The manufacture of L–NH_3 from natural gas reforming without carbon capture and storage (CCS) yields 2.24 kg of CO_2-equivalent emissions per kilogram of L–NH_3. Nonetheless, the implementation of CCS technology decreases GHG emissions by 24%. Upon evaluating the complete life cycle, the total emissions of L–NH_3 from natural gas reforming amount to 1.87 kg CO_2-eq per kilogram of L–NH_3, rendering it a lower-emission fuel in comparison to LNG, DME, and methanol.

To adhere to environmental legislation and conform to United Nations Environment Programme (UNEP) standards, multiple recommendations are suggested. Considering Qatar's large natural gas reserves, the production and exportation of alternative fuels—such as L–NH_3, methanol, and DME—rather than an exclusive emphasis on LNG, could alleviate international pressure on environmental policies and establish Qatar as a responsible energy supplier in terms of environmental stewardship. Storing natural gas as L–NH_3, methanol, or DME rather than LNG can reduce overall greenhouse gas emissions, as LNG storage incurs persistent boil-off gas losses, resulting in elevated emissions over time. Enhancing the use of ammonia and hydrogen as energy carriers offers obvious environmental advantages, as both fuels emit no CO_2 during combustion and may be included into low-carbon energy systems. The production and transportation of methanol and DME yield fewer GHG emissions than LNG; however, this benefit is negated when their use in internal combustion engines is evaluated. Consequently, deployment strategies must prioritize the optimization of fuel use in applications that enhance emission reduction advantages.

This study offers a thorough assessment of life cycle GHG emissions; yet, future research should incorporate other environmental impact categories. Broadening the analysis to include ozone layer depletion, acidification, and marine sediment ecotoxicity will yield a more comprehensive environmental viewpoint on various energy carriers. Given that BOG emissions differ in type and quantity based on fuel characteristics, examining these supplementary effect categories will provide a more comprehensive environmental evaluation. Moreover, examining the capture and reutilization of BOG in combustion applications during transportation can improve the environmental efficacy of these energy carriers, hence diminishing emissions.

This study emphasizes the necessity of implementing cleaner energy production techniques, enhancing fuel storage and transportation ways, and incorporating renewable electricity sources to mitigate the environmental impact of LNG, DME, methanol, L–H_2, and L–NH_3. Through the diversification of energy carrier production, the utilization of carbon capture technology, and the enhancement of renewable energy's role, Qatar may fortify its status as a global leader in sustainable energy production and exportation, while adhering to international climate obligations.

References

1. Tagliaferri C, Clift R, Lettieri P, Chapman C (2017) Liquefied natural gas for the UK: a life cycle assessment. Int J Life Cycle Assess 22:1944–1956. https://doi.org/10.1007/s11367-017-1285-z
2. Noor Akashah MH, Mohammad Rozali NE, Mahadzir S, Liew PY (2023) Utilization of cold energy from LNG regasification process: a review of current trends. Processes 11. https://doi.org/10.3390/pr11020517
3. Buttler A, Spliethoff H (2018) Current status of water electrolysis for energy storage, grid balancing and sector coupling via power-to-gas and power-to-liquids: a review. Renew Sustain Energy Rev 82:2440–2454. https://doi.org/10.1016/j.rser.2017.09.003
4. Halder P, Babaie M, Salek F, Haque N, Savage R, Stevanovic S et al (2024) Advancements in hydrogen production, storage, distribution and refuelling for a sustainable transport sector: hydrogen fuel cell vehicles. Int J Hydrogen Energy 52:973–1004. https://doi.org/10.1016/j.ijhydene.2023.07.204
5. Wen D, Aziz M (2022) Techno-economic analyses of power-to-ammonia-to-power and biomass-to-ammonia-to-power pathways for carbon neutrality scenario. Appl Energy 319. https://doi.org/10.1016/j.apenergy.2022.119272
6. Valera-Medina A, Xiao H, Owen-Jones M, David WIF, Bowen PJ (2018) Ammonia for power. Prog Energy Combust Sci 69:63–102. https://doi.org/10.1016/j.pecs.2018.07.001
7. Olah GA (2005) Beyond oil and gas: the methanol economy. Angew Chemie Int Ed 44:2636–2639. https://doi.org/10.1002/anie.200462121
8. Kotowicz J, Brzęczek M, Walewska A, Szykowska K (2022) Methanol production in the Brayton cycle. Energies 15. https://doi.org/10.3390/en15041480
9. Soltic P, Hilfiker T, Wright Y, Hardy G, Fröhlich B, Klein D (2024) The potential of dimethyl ether (DME) to meet current and future emissions standards in heavy-duty compression-ignition engines. Fuel 355:129357. https://doi.org/10.1016/j.fuel.2023.129357
10. Singh T, Jalwal S, Chakraborty S (2022) Homogeneous first-row transition-metal-catalyzed carbon dioxide hydrogenation to formic acid/formate, and methanol. Asian J Org Chem 11. https://doi.org/10.1002/ajoc.202200330
11. Al-Breiki M, Bicer Y (2020) Technical assessment of liquefied natural gas, ammonia and methanol for overseas energy transport based on energy and exergy analyses. Int J Hydrog Energy. https://doi.org/10.1016/j.ijhydene.2020.04.181
12. Al-Breiki M, Bicer Y (2020) Investigating the technical feasibility of various energy carriers for alternative and sustainable overseas energy transport scenarios. Energy Convers Manag 209:112652. https://doi.org/10.1016/j.enconman.2020.112652

13. Dincer I, Rosen MA (2013) Exergy analysis of countries, regions, and economic sectors. Exergy Elsevier, pp 425–450. https://doi.org/10.1016/B978-0-08-097089-9.00021-8
14. Moran MJ, Shapiro HN (1993) Fundamentals of engineering thermodynamics, second edition. Eur J Eng Educ 18:215. https://doi.org/10.1080/03043799308928176
15. Klein (2011) EES: engineering equation solver | F-chart software: Engineering Software
16. Morgan E (2013) Techno-economic feasibility study of ammonia plants powered by offshore wind. Open Access Dis. https://doi.org/10.7275/11kt-3f59
17. Morgan E, Manwell J, McGowan J (2014) Wind-powered ammonia fuel production for remote islands: a case study. Renew Energy 72:51–61. https://doi.org/10.1016/j.renene.2014.06.034
18. Mansour K (2012) Fires in large atmospheric storage tanks and their effect on adjacent tanks. Dr Thesis, p 383
19. Qatargas (2016) Sustainability report 2016. Doha
20. Qatar Fertiliser Company (QAFCO) (2015) Sustainability report 2015
21. Mannion K (2018) Towards a low carbon future, vol 240. https://doi.org/10.1016/S0262-4079(18)31934-1
22. Kurle YM, Wang S, Xu Q (2017) Dynamic simulation of LNG loading, BOG generation, and BOG recovery at LNG exporting terminals. Comput Chem Eng 97:47–58. https://doi.org/10.1016/j.compchemeng.2016.11.006
23. Dobrota Đ, Lalić B, Komar I (2013) Problem of Boil-off in LNG supply chain. Trans Marit Sci 2:91–100. https://doi.org/10.7225/toms.v02.n02.001
24. Liu C, Zhang J, Xu Q, Gossage JL (2010) Thermodynamic-analysis-based design and operation for boil-off gas flare minimization at LNG receiving terminals. Ind Eng Chem Res 49:7412–7420. https://doi.org/10.1021/ie1008426
25. Adamkiewicz A, Cydejko J (2016) The influence of energy consumption of gas vapour reliquefaction on the structure of the LNG carrier power plant. Naše More 63:38–42. https://doi.org/10.17818/NM/2016/1.6
26. Pospíšil J, Charvát P, Arsenyeva O, Klimeš L, Špiláček M, Klemeš JJ (2019) Energy demand of liquefaction and regasification of natural gas and the potential of LNG for operative thermal energy storage. Renew Sustain Energy Rev 99:1–15. https://doi.org/10.1016/j.rser.2018.09.027
27. Shin MW, Shin D, Choi SH, Yoon ES, Han C (2007) Optimization of the operation of boil-off gas compressors at a liquified natural gas gasification plant. Ind Eng Chem Res 46:6540–6545. https://doi.org/10.1021/ie061264i
28. Hasan MMF, Zheng AM, Karimi IA (2009) Minimizing boil-off losses in liquefied natural gas transportation. Ind Eng Chem Res 48:9571–9580. https://doi.org/10.1021/ie801975q
29. Morgan ER (2013) Techno-economic feasibility study of ammonia plants powered by offshore wind. University of Massachusetts—Amherst, PhD Dissertation, p 432
30. Molnar G (2022) Economics of gas transportation by pipeline and LNG. Palgrave Handb Int Energy Econ 23–57. https://doi.org/10.1007/978-3-030-86884-0_2
31. EIA. Short-Term Energy Outlook (2023) Adm US Energy Inf. https://www.eia.gov/todayinenergy/detail.php?id=61183
32. Connelly E, Michael Penev AE, Hunter C (2019) Current status of hydrogen liquefaction costs. Dep Energy Hydrog Fuel Cells Progr 2019:1–10
33. Jabarivelisdeh B, Jin E, Christopher P, Masanet E. (2022) Ammonia production processes from energy and emissions perspectives: a technical brief
34. Methanex (2022) The global methanol leader. Geismar
35. Park SH, Lee CS (2014) Applicability of dimethyl ether (DME) in a compression ignition engine as an alternative fuel. Energy Convers Manag 86:848–863. https://doi.org/10.1016/j.enconman.2014.06.051
36. Somoza-Tornos A, Guerra OJ, Crow AM, Smith WA, Hodge BM (2021) Process modeling, techno-economic assessment, and life cycle assessment of the electrochemical reduction of CO_2: a review. iScience 24. https://doi.org/10.1016/j.isci.2021.102813
37. Al-Breiki M, Bicer Y (2020) Comparative cost assessment of sustainable energy carriers produced from natural gas accounting for boil-off gas and social cost of carbon. Energy Rep 6:1897–1909. https://doi.org/10.1016/j.egyr.2020.07.013

38. QatarGas (2020) North field. Qatargas Oper Co Ltd. http://www.qatargas.com/english/aboutus/north-field
39. Qatargas (2018) Qatargas' chartered fleet. Qatargas Oper Co Ltd. http://www.qatargas.com/english/operations/qatargas-chartered-fleet
40. Moon JW, Lee YP, Jin YW, Hong ES (2007) Cryogenic refrigeration cycle for re-liquefaction of LNG boil-off gas. Cryocoolers 14:629–635
41. Gómez JR, Gómez MR, Garcia RF, Catoira ADM (2013) On board LNG reliquefaction technology: a comparative study. Polish Marit Res 21:77–88. https://doi.org/10.2478/pomr-2014-0011
42. Software F-Chart (2015) EES: engineering equation solver: engineering software
43. Aj F, Gl J, Tr T (1999) Developments in natural gas liquefaction. Hydrocarb Process 78:47–56
44. Seddon DD (2006) Gas usage and value: the technology and economics of natural gas use in the process industries. PennWell Corporation, Tulsa. 1-59370-073-3
45. Dahl CA (2006) Book review—international energy markets: understanding pricing, policies, and profits, vol 27. PennWell Corporation. https://doi.org/10.5547/issn0195-6574-ej-vol27-no1-9
46. Schnitkey G (2018) Nitrogen prices, rates cuts, and 2018 fertilizer costs. Farmdoc Dly (8)58 (Dep Agric Consum Econ Univ Illinois Urbana-Champaign Apr 3 2018). https://farmdocdaily.illinois.edu/2018/04/nitrogen-prices-rates-cuts-and-2018-fertilizer-cost.html
47. Methanex (2015) Methanex methanol price sheet. METHANEX Corp, p 1. https://www.methanex.com/sites/default/files/methanol-price/Mx-Price-Sheet-2015-11-27.pdf
48. CEIC (2018) China CN: market price: monthly avg: organic chemical material: dimethyl ether: 99.0% or above. Ceic. https://www.ceicdata.com/en/china/china-petroleum--chemical-industry-association-petrochemical-price-organic-chemical-material/cn-market-price-monthly-avg-organic-chemical-material-dimethyl-ether-990-or-above
49. ARENA (2018) Opportunities for Australia from hydrogen exports Acil Allen consulting for Arena. Melbourne
50. Seddon DD (2006) Gas usage and value: the technology and economics of natural gas use in the process industries
51. Kamalinejad M, Sheykhbahaee A, Mazaheri S (2016) Financial feasibility study between purchasing and hiring LNG carrier In Iranian LNG Industry, p 1
52. Magraw K (2017) I.32 United Nations conference on trade and development (UNCTAD). In: The Elgar encyclopedia of international economic law. Edward Elgar Publishing, pp 73–76. https://doi.org/10.4337/9781784713546.45
53. Engines T (2013) Propulsion trends in LNG carriers
54. Sharples J (2019) LNG supply chains and the development of LNG as a shipping fuel in Northern Europe. https://doi.org/10.26889/9781784671266
55. PortWorld (2012) PortWorld distance—ship Voyage distance calculator. http://www.portworld.com/map/
56. Singapore Bunker Prices—Ship & Bunker (2020) Sh Bunker. https://shipandbunker.com/prices/apac/sea/sg-sin-singapore#IFO180
57. Eveloy V, Gebreegziabher T (2018) A review of projected power-to-gas deployment scenarios. Energies 11. https://doi.org/10.3390/en11071824
58. Xu W, Shah Z, Solangi W et al (2019) Economic viability and environmental efficiency analysis of hydrogen production processes for the decarbonization of energy systems. Processes 7:494. https://doi.org/10.3390/pr7080494
59. Al-Breiki M, Bicer Y (2020) Comparative life cycle assessment of sustainable energy carriers including production, storage, overseas transport and utilization. J Clean Prod
60. International Organization for Standardization (2006) Environmental management—life cycle assessment—principles and framework. Lausanne
61. Curran MA (2008) Life-cycle assessment. In: Encyclopedia of ecology, five volume set. Elsevier, pp 2168–2174. https://doi.org/10.1016/B978-008045405-4.00629-7

62. Korre A, Nie Z, Durucan S (2012) Life cycle assessment of the natural gas supply chain and power generation options with CO_2 capture and storage: assessment of Qatar natural gas production, LNG transport and power generation in the UK. Sustain Technol Syst Policies 2012:11. https://doi.org/10.5339/stsp.2012.ccs.11
63. Herva M, Franco A, Carrasco EF, Roca E (2011) Review of corporate environmental indicators. J Clean Prod 19:1687–1699. https://doi.org/10.1016/j.jclepro.2011.05.019
64. Argonne National Laboratory (2012) Argonne GREET model webpage
65. IPCC (2013) IPCC fifth assessment report: CSIROexperts comment. https://doi.org/10.1071/ec13228
66. Belton V, Stewart TJ (2002) Multiple criteria decision analysis. Springer US, Boston, MA. https://doi.org/10.1007/978-1-4615-1495-4
67. Arteconi A, Brandoni C, Evangelista D, Polonara F (2010) Life-cycle greenhouse gas analysis of LNG as a heavy vehicle fuel in Europe. Appl Energy 87:2005–2013. https://doi.org/10.1016/j.apenergy.2009.11.012
68. Bhandari R, Trudewind CA, Zap P (2012) Life cycle assessment of hydrogen production methods—a review
69. Derwent R, Simmonds P, O'Doherty S, Manning A, Collins W, Stevenson D (2006) Global environmental impacts of the hydrogen economy. Int J Nucl Hydrog Prod Appl 1:57. https://doi.org/10.1504/ijnhpa.2006.009869
70. Ashida Y, Arashiba K, Nakajima K, Nishibayashi Y (2019) Molybdenum-catalysed ammonia production with samarium diiodide and alcohols or water. Nature 568:536–540. https://doi.org/10.1038/s41586-019-1134-2
71. Mehmeti A, Angelis-Dimakis A, Arampatzis G, McPhail S, Ulgiati S (2018) Life cycle assessment and water footprint of hydrogen production methods: from conventional to emerging technologies. Environments 5:24. https://doi.org/10.3390/environments5020024
72. Bicer Y, Dincer I, Zamfirescu C, Vezina G, Raso F (2016) Comparative life cycle assessment of various ammonia production methods. J Clean Prod 135:1379–1395. https://doi.org/10.1016/j.jclepro.2016.07.023
73. Amponsah NY, Troldborg M, Kington B, Aalders I, Hough RL (2014) Greenhouse gas emissions from renewable energy sources: a review of lifecycle considerations. Renew Sustain Energy Rev 39:461–475. https://doi.org/10.1016/j.rser.2014.07.087
74. Ozawa A, Kudoh Y, Kitagawa N, Muramatsu R (2019) Life cycle CO_2 emissions from power generation using hydrogen energy carriers. Int J Hydrogen Energy 44:11219–11232. https://doi.org/10.1016/j.ijhydene.2019.02.230
75. Bicer Y, Dincer I (2017) Assessment of a sustainable electrochemical ammonia production system using photoelectrochemically produced hydrogen under concentrated sunlight. ACS Sustain Chem Eng 5:8035–8043. https://doi.org/10.1021/acssuschemeng.7b01638
76. Solomon DS (2017) Investigation of production of dimethyl ether (DME) from renewable resources and its integration into the oil production system
77. John P (2020) Qatar accounts for bulk of LNG supplies to Asia in 2019: IGU. Gulf Times
78. Oeder S, Kanashova T, Sippula O, Sapcariu SC, Streibel T, Arteaga-Salas JM et al (2015) Particulate matter from both heavy fuel oil and diesel fuel shipping emissions show strong biological effects on human lung cells at realistic and comparable in vitro exposure conditions. PLoS ONE 10:e0126536. https://doi.org/10.1371/journal.pone.0126536
79. Kandijk G (2015) Environmental and economic aspects of using LNG as a fuel for shipping in The Netherlands
80. DNV GL—Maritime (2019) Assessment of selected alternative fuels and technologies, vol 391
81. Edwards R, Larive J-F, Mahieu V, Rounveirolles P (2007) Well-to-wheels analysis of future automotive fuels and well-to-wheels report, vol version 2c. https://doi.org/10.2788/79018
82. Aronoff E, Sitty G, Taft N (2014) The opportunities for emission reduction of methanol fuel made from natural gas Eyal Aronoff, Gal Sitty, Nathan Taft. In: The opportunities for emission reduction of methanol fuel made from natural gas

References

83. Semelsberger TA, Borup RL, Greene HL (2006) Dimethyl ether (DME) as an alternative fuel. J Power Sour 156:497–511. https://doi.org/10.1016/j.jpowsour.2005.05.082
84. Brynolf S, Fridell E, Andersson K (2014) Environmental assessment of marine fuels: liquefied natural gas, liquefied biogas, methanol and bio-methanol. J Clean Prod 74:86–95. https://doi.org/10.1016/j.jclepro.2014.03.052
85. Higo M, Dowaki K (2010) A life cycle analysis on a bio-DME production system considering the species of biomass feedstock in Japan and Papua New Guinea. Appl Energy 87:58–67. https://doi.org/10.1016/j.apenergy.2009.08.030
86. Baptista P, Ribau J, Bravo J, Silva C, Adcock P, Kells A (2011) Fuel cell hybrid taxi life cycle analysis. Energy Policy 39:4683–4691. https://doi.org/10.1016/j.enpol.2011.06.064
87. Kongshaug G, Jenssen TK (2003) Energy consumption and greenhouse gas emissions in fertilizer production
88. Spoof-Tuomi K, Niemi S (2020) Environmental and economic evaluation of fuel choices for short sea shipping. Clean Technol 2:34–52. https://doi.org/10.3390/cleantechnol2010004
89. Wood S, Cowie A (2004) A review of greenhouse gas emission factors for fertiliser production. IEA Bioenergy Task 38:20
90. Peng T, Zhou S, Yuan Z, Ou X (2017) Life cycle greenhouse gas analysis of multiple vehicle fuel pathways in China. Sustain 9. https://doi.org/10.3390/su9122183
91. Pehnt M (2000) Life cycle assessment of fuel cells and relevant fuel chains, vol 11
92. Peng T, Zhou S, Yuan Z, Ou X (2017) Life cycle greenhouse gas analysis of multiple vehicle fuel pathways in China. Sustain 9:2183. https://doi.org/10.3390/su9122183
93. Lee U, Han J, Wang M, Ward J, Hicks E, Goodwin D et al (2016) Well-to-wheels emissions of greenhouse gases and air pollutants of dimethyl ether from natural gas and renewable feedstocks in comparison with petroleum gasoline and diesel in the United States and Europe. SAE Int J Fuels Lubr 9:546–557. https://doi.org/10.4271/2016-01-2209
94. Wang M (1999) Technical report: GREET 1.5—transportation fuel-cycle model—vol 1: methodology, development, use, and results. Argonne
95. Lewis J (2018) Fuels without carbon: prospects and the pathway forward for zero-carbon hydrogen and ammonia fuel. Ammon Energy Assoc. https://www.ammoniaenergy.org/articles/fuels-without-carbon-prospects-and-the-pathway-forward-for-zero-carbon-hydrogen-and-ammonia-fuels/
96. Brohi EA (2014) Ammonia as fuel for internal combustion engines?
97. Bengtsson S, Andersson K, Fridell E (2013) An environmental life cycle assessment of LNG and HFO as marine fuels. Environ Life Cycle Assess LNG HFO Mar Fuels 225:97–110. https://doi.org/10.1177/1475090211402136
98. Elgowainy A, Mintz M, Steward D, Antonia O, Brown D, Gardiner M (2012) Liquid hydrogen production and delivery from a dedicated wind power plant. https://doi.org/10.1016/S1474-4422(12)70173-4
99. Bicer Y, Khalid F, Mohamed AMO, Al-Breiki M, Ali MM (2020) Electrochemical modelling of ammonia synthesis in molten salt medium for renewable fuel production using wind power. Int J Hydrogen Energy. https://doi.org/10.1016/j.ijhydene.2020.03.085
100. IMO (2009) Technical information on systems and operation to assist development of VOC management plan

Open Access This chapter is licensed under the terms of the Creative Commons Attribution 4.0 International License (http://creativecommons.org/licenses/by/4.0/), which permits use, sharing, adaptation, distribution and reproduction in any medium or format, as long as you give appropriate credit to the original author(s) and the source, provide a link to the Creative Commons license and indicate if changes were made.

The images or other third party material in this chapter are included in the chapter's Creative Commons license, unless indicated otherwise in a credit line to the material. If material is not included in the chapter's Creative Commons license and your intended use is not permitted by statutory regulation or exceeds the permitted use, you will need to obtain permission directly from the copyright holder.

Chapter 9
Future Directions

9.1 Future Energy Systems

9.1.1 Decentralized and Distributed Energy Systems

A substantial transformation is occurring in the conventional centralized energy paradigm, which is defined by the transmission of electricity over long distances by large power plants. This transition is propelled by progress in renewable energy technologies, the rising aspiration for energy resilience and self-reliance, and the escalating need for increased consumer involvement in the energy market. The implementation of decentralized and distributed energy systems (DEDS) holds great potential in achieving a sustainable, resilient, and equitable energy future.

Microgrids, which are decentralized power networks that can function autonomously or in collaboration with the main grid, are a fundamental element of the DEDS movement. Community and organizational energy systems facilitate the production, distribution, and consumption of energy at a local level, so decreasing reliance on the centralized grid and enhancing energy resilience in the presence of disturbances. Typically constructed around microgrids, community energy systems enable individuals and groups to assume responsibility for their energy generation and consumption. By consolidating resources and allocating funds to renewable energy infrastructure, communities can diminish their carbon emissions, establish local employment opportunities, and generate supplementary funding through energy sales. An exemplary case is the Brooklyn Microgrid project, a community-driven endeavor in New York City that seeks to establish a decentralized energy trading system for electricity produced from solar sources [1].

Peer-to-peer (P2P) energy trading platforms, facilitated by blockchain technology and smart grid infrastructure, are fundamentally transforming the process of energy exchange. Platforms of this nature facilitate the direct exchange of energy between prosumers, who are individuals or entities engaged in both energy production and consumption, so circumventing conventional energy retailers and utilities. Peer-to-peer energy trading facilitates increased consumer options, improves market economic efficiency, and stimulates the uptake of renewable energy technologies. The blockchain-based peer-to-peer energy trading platform developed by Power Ledger, an Australian company, is currently undergoing a pilot phase in multiple countries, including Australia, Thailand, and the United States [2].

The decentralized structure of DEDS renders them highly suitable for the incorporation of renewable energy sources, such as solar and wind, which normally exhibit their intermittent and dispersed characteristics. The implementation of DEDS at the local level can effectively mitigate transmission losses, improve grid stability, and expedite the shift towards a sustainable energy future. The "Energiewende" (energy transition) regime in Germany, which seeks to attain a significant proportion of renewable energy in the nation's energy composition, has stimulated the expansion of several community-owned wind and solar initiatives [3].

Future energy systems are defined by decentralization, distribution, and a growing dependence on renewable sources. Emerging technologies such as microgrids, community energy systems, P2P energy trading, and the incorporation of renewable energy at the local level are revolutionizing the energy sector and generating fresh prospects for individuals, communities, and businesses to engage in a more sustainable and fair energy future.

9.1.2 Advanced Energy Storage Technologies

Robust and efficient energy storage solutions are required to guarantee a dependable and stable power supply due to the heterogeneity and intermittent nature of renewable energy sources. Technologically advanced energy storage systems are essential in facilitating the shift towards a more environmentally friendly and enduring energy future.

Solid-state batteries, characterized by their solid electrolytes as opposed to the liquid or gel formulations used in traditional lithium-ion batteries, offer substantial progress in energy storage. These batteries provide increased energy density, enhanced safety, and extended lifespans, rendering them a compelling choice for electric vehicles and large-scale energy storage systems. Specifically, Toyota is making significant investments in the advancement of solid-state batteries with the goal of making them available for electric vehicles [4].

Hydrogen fuel cells provide a flexible and environmentally friendly means of storing energy by transforming hydrogen and oxygen into electrochemical energy, with water being the sole byproduct. They have a wide range of applications, encompassing transportation, stationary power production, and portable power transmission. The viability of hydrogen fuel cell technology for sustainable transportation has been demonstrated by Hyundai, a prominent automotive manufacturer, through the development of hydrogen fuel cell vehicles, such as the NEXO SUV [5].

In concentrated solar power plants and industrial processes, TES technologies, such as molten salt and phase change materials, are employed to store thermal energy for subsequent utilization, so enhancing operational efficiency. Power pumped hydro and compressed air energy storage are mechanical energy storage technologies that store energy as potential or kinetic energy, enabling grid-scale energy storage and supplementary services. The Hornsdale Power Reserve in Australia, which hosts the largest lithium-ion battery globally, serves as evidence of the increasing significance of sophisticated energy storage in stabilizing power grids and enabling the incorporation of renewable energy sources [6].

The utilization of advanced energy storage technologies is essential for fully harnessing the capabilities of renewable energy and constructing a more sustainable and robust energy future. Solid-state batteries, hydrogen fuel cells, thermal and mechanical energy storage systems, and other developing technologies are facilitating the transition towards a sustainable energy sector.

9.1.3 Digitalization and Smart Grids

The adoption of digital technologies, including the Internet of Things (IoT), artificial intelligence (AI), and advanced analytics, is revolutionizing the energy industry, resulting in the development of intelligent power networks. Smart grids are distinguished by their capacity to gather, evaluate, and react to data in real time, allowing for efficient energy management, increased reliable operation of the grid, and improved interaction with consumers.

The IoT, a network of linked device and sensor systems, facilitates the gathering of extensive quantities of data regarding energy production, transmission, distribution, and consumption. The integration of this data with AI algorithms enables the prediction of energy demand, optimization of energy flows, detection and response to grid anomalies, and facilitation of the integration of renewable energy sources. The Nest thermostats employ machine learning algorithms to acquire knowledge of user preferences and enhance energy efficiency in residential settings [7].

Utilizing artificial intelligence and machine learning, predictive analytics can accurately predict energy demand, generation, and imminent grid disruptions, so facilitating proactive management and optimization of the grid. Forecasting future events and trends enables grid operators to foresee and alleviate possible problems, such as power outages or equipment malfunctions, resulting in enhanced grid dependability and resilience. Siemens' EnergyIP platform employs predictive analytics to enhance grid operations and maintenance, resulting in cost reduction and overall asset performance improvement [8].

The growing dependence on digital technologies in the energy industry gives rise to apprehensions regarding weaknesses in cybersecurity. The inherent interconnectivity of smart grids, coupled with the increasing amount of sensitive data being gathered and transmitted, presents possible vulnerabilities for cyberattacks.

To guarantee the cybersecurity of future energy systems, it is necessary to implement strong security measures, maintain constant monitoring, and adopt proactive strategies to mitigate threats. The 2021 ransomware attack on the Colonial Pipeline network exposed the susceptibility of vital energy infrastructure to cyberattacks and emphasized the need of strong cybersecurity measures [9].

The process of digitalization and the rise of smart grids are revolutionizing the energy sector, therefore facilitating enhanced efficiency, dependability, and customer empowerment. The incorporation of IoT, AI, and predictive analytics is enhancing energy management and grid operations, while cybersecurity continues to be a crucial obstacle that demands continuous focus and financial resources.

9.1.4 Renewable Energy Integration and Hybrid Systems

The inherent unpredictability of renewable energy sources, such as solar and wind, has historically presented obstacles to their incorporation into the electrical grid. Nevertheless, progress in energy storage technologies and the emergence of hybrid renewable systems are facilitating a future in which renewables can consistently supply reliable baseload power and expedite the energy transition.

The integration of solar, wind, and energy storage technologies generates a synergistic impact, facilitating a dependable and consistent power provision. Solar and wind energy generation are mutually supplemented, with solar energy generating more during daylight hours and wind energy often generating more during nighttime. The storage of excess energy during periods of high generation and subsequent release during periods of low generation by energy storage systems, such as batteries or pumped hydro, guarantees a steady and equitable power supply. Initiatives such as the Kennedy Energy Park in Australia, which integrate solar, wind, and battery storage, are showcasing the practicality of renewable energy hybrid systems in delivering dependable baseload electricity [10].

Hybrid renewable systems provide a versatile and adjustable solution to address a wide range of energy requirements, especially in isolated or off-grid areas. Through the integration of several renewable energy sources and energy storage technologies, these systems offer a dependable and economically viable substitute for conventional power production produced from fossil fuels. The incorporation of hybrid systems can expedite the energy transition by facilitating increased integration of renewable energy into the power grid, curbing greenhouse gas emissions, and fostering energy self-sufficiency. Located in El Hierro, Canary Islands, the Gorona del Viento project is an innovative hybrid renewable energy system that integrates wind and pumped hydro storage to provide electricity for the entire island [11].

The integration of renewable energy sources and the development of hybrid systems are the key factors responsible for shaping the future of energy. The integration of solar, wind, and energy storage technologies is facilitating the development of a future in which renewable energy can consistently supply baseload power. By embracing hybrid renewable systems, the energy transition can be expedited, so resulting in a more environmentally friendly, sustainable, and robust energy future.

9.1.5 Carbon Capture, Utilization, and Storage

Although the shift towards renewable energy sources is of utmost importance, the presence of current fossil fuel infrastructure and industries that are resistant to change compels the creation and implementation of technologies aimed at reducing carbon emissions. CCUS technologies provide a mechanism to reduce carbon emissions in the energy industry and reach the goal of achieving net-zero emissions.

Post-combustion capture, a conventional method of carbon capture, is characterized by its high energy consumption and potential high costs. Next-generation technologies, such as direct air capture (DAC) and membrane-based capture, have the potential to enhance efficiency and reduce costs. DAC technologies have the capability to directly capture carbon dioxide from the atmosphere, offering a versatile approach for carbon emission reduction. Climeworks, a Swiss firm, is leading the way in commercializing DAC technology. Their Orca facility in Iceland is the first to capture and store atmospheric carbon dioxide [12].

Integrating CCUS with future energy systems is essential for effectively achieving profound decarbonization. CCUS can be integrated with fossil fuel power plants to mitigate emissions, or it can be employed alongside renewable energy sources, such as biomass, to develop technologies that generate negative emissions. Moreover, effectively captured carbon dioxide can be employed as a raw material for diverse industrial operations, including the synthesis of synthetic fuels or construction materials. The Boundary Dam Power Station in Canada is the planet's inaugural large-scale coal-fired power plant equipped with CCUS, showcasing the capacity of this technology to reduce carbon emissions from current energy infrastructure [13].

CCUS technologies are crucial in reducing carbon emissions and attaining net-zero objectives. Next-generation technologies, such as DAC and membrane-based capture, provide potential avenues to enhance effectiveness and lower the expenses of carbon capture. By integrating CCUS with forthcoming energy systems, such as renewable energy and industrial processes, it is possible to achieve profound decarbonization and establish a sustainable energy future.

9.2 Future Sustainable Fuel Market

9.2.1 Market Trends for Hydrogen and Ammonia

The shift towards a sustainable energy future is fueling the rise of novel energy carriers, with hydrogen and ammonia seizing considerable prominence. These multifunctional fuels have the capacity to reduce carbon emissions in several sectors, such as transportation, industry, and power production. The hydrogen and ammonia market is expected to experience substantial expansion due to technical progress, favorable environmental regulations, and rising need for sustainable energy alternatives.

The projected period for the next decades is expected to see significant expansion in the worldwide demand for hydrogen and ammonia. By 2030, the IEA predicts that hydrogen demand may reach 180 million tons, mostly due to its application in transportation, industry, and buildings [3]. Due to its role as a hydrogen carrier, ammonia is projected to experience substantial demand growth, with estimates suggesting about USD 14 billion in annual capital investment for ammonia production to 2050 [14].

The utilization of hydrogen and ammonia is expanding in several sectors due to their capacity to mitigate carbon emissions and improve energy security. In the transportation industry, hydrogen fuel cell vehicles and ammonia-powered ships are becoming as viable substitutes for traditional vehicles powered by fossil fuels. Furthermore, the industrial sector is investigating the utilization of hydrogen and ammonia as raw materials for diverse operations, including steel production and fertilizer manufacturing. In addition, the power industry is exploring the application of hydrogen and ammonia for the purposes of energy storage and grid balancing. An example of this is Saudi Arabia's ongoing development of a mega-city named NEOM, which is exclusively powered by renewable energy and green hydrogen. The objective of this project is to showcase the potential of hydrogen as a crucial energy carrier for the future [15].

Despite the promising growth prospects of the hydrogen and ammonia market, there are several obstacles that must be overcome in order to fully exploit its potential. The establishment of sufficient infrastructure, encompassing manufacturing facilities, transportation networks, and storage capacities, is essential to enable the extensive widespread use of these fuels. Furthermore, it is crucial to mitigate the production costs and enhance the efficiency of hydrogen and ammonia technologies in order to bolster their competitiveness against traditional fuels. Effective collaboration between policymakers and industry stakeholders is necessary to establish a conducive environment for the growth of a strong hydrogen and ammonia commercial sector.

Sustainable fuels' future hinges on the advancement and widespread use of hydrogen and ammonia. The projected global demand for these fuels is expected to experience substantial growth, primarily due to their capacity to reduce carbon emissions in different industries and improve energy security. Notwithstanding the existing obstacles in infrastructure development and cost reduction, the market for hydrogen and ammonia is positioned for a revolutionary growth, which will contribute to a more environmentally friendly and sustainable energy future.

9.2.2 Policy and Regulatory Drivers

The shift towards a sustainable fuel market is driven by both market dynamics and technical progress, as well as a favorable policy and regulatory structure. Governments and international organizations are exerting significant influence on the trajectory of sustainable fuels by employing diverse strategies such as carbon pricing, renewable energy objectives, and incentives to promote the use of sustainable fuels.

Carbon pricing mechanisms, such as carbon taxes or emissions trading schemes (ETS), implement a financial burden on carbon emissions, so motivating both industries and consumers to decrease their carbon footprint and transition to more environmentally friendly energy technologies. The European Union Emissions Trading System (EU ETS), which is the largest carbon market globally, has proven to be highly successful in promoting emissions reductions through the implementation of carbon pricing [16]. The internalization of carbon emissions costs by these mechanisms establishes equitable conditions for the use of sustainable fuels and promotes their widespread acceptance.

Diplomatic accords, such as the Paris Agreement on climate change, establish ambitious objectives for decreasing greenhouse gas emissions and encouraging the adoption of sustainable energy sources. These agreements establish a worldwide structure for taking action and promote countries to implement policies that facilitate the advancement and implementation of sustainable fuels. Additionally, numerous nations have established national renewable energy goals, which further stimulate the demand for sustainable fuels and encourage investments in clean energy technologies.

A range of incentives, including tax exemptions, subsidies, and loan guarantees, are being implemented by national governments to encourage the use of sustainable fuels. The purpose of these incentives is to decrease the price of environmentally friendly fuels and promote their use in markets such as transportation, industry, and power production. For instance, the United States provides tax incentives for the manufacturing and utilization of renewable fuels, such as hydrogen and fuels derived from biomass [17]. The provision of such incentives is of utmost importance in establishing a conducive market climate for sustainable fuels and expediting their implementation.

Policy and regulatory factors play a crucial role in determining the trajectory of sustainable fuels. Carbon pricing mechanisms, combined with international agreements and national incentives, establish a conducive framework for the advancement and acceptance of clean energy technologies. Through the provision of unambiguous signals to the market and the promotion of investments in sustainable fuels, these mechanisms play a crucial role in expediting the shift towards a more environmentally friendly and sustainable energy future.

9.2.3 Investment and Financing in Sustainable Fuels

Substantial investments in research, development, and infrastructure are necessary for the advancement and implementation of sustainable fuels. Both the public and private sectors are essential in facilitating the mobilization of capital and promoting innovation in this particular industry.

An array of investors, encompassing venture capitalists, private equity firms, and institutional investors, are actively allocating funds towards sustainable fuel innovations. The appealing growth prospects, together with the growing focus on environmental, social, and governance (ESG) aspects, are attracting substantial investment into this industry. A range of financial models, such as project finance, equity investments, and public–private partnerships, are being used to finance the advancement and commercialization of sustainable fuel technologies. Hy24, a collaborative effort between Ardian and FiveT Hydrogen, is a specialized investment platform focused on clean hydrogen infrastructure. It has secured more than €2 billion in committed capital [18].

The acceleration of sustainable fuel development and deployment is contingent upon the essential collaboration between the public and private sectors. Governments can implement supportive measures, such as adopting carbon pricing, setting renewable energy targets, and offering incentives for the use of sustainable fuels, in order to establish a conducive market environment. Conversely, the private sector has the ability to utilize its knowledge and resources to create groundbreaking technologies and introduce them to the industrial market. Public–private collaborations can optimize the distribution of risks and benefits, so promoting the effective execution of extensive sustainable fuel initiatives. A noteworthy instance of such cooperation is the European Clean Hydrogen Alliance, which unites industry, investors, and policymakers to expedite the establishment of a hydrogen ecosystem in Europe [16].

Capital investment and funding are essential for the advancement and implementation of sustainable fuels. A wide array of investors and financial models are being used to facilitate the expansion of this industry. Effective cooperation between the public and private sectors is crucial in establishing a conducive atmosphere for sustainable fuel initiatives and expediting the shift towards a more environmentally friendly and sustainable energy future.

9.2.4 Future Market Challenges

Although sustainable fuels show promise for the future, they encounter several obstacles that must be resolved in order to attain broad acceptance and commercial viability. The challenges include competition with current fuels, scalability and supply chain deficiencies, and surmounting technological and economic obstacles.

At their early stages of development, sustainable fuels encounter intense competition from well-established and deeply rooted fossil fuels. Contemporary infrastructure, supply networks, and consumer preferences are strongly biased towards conventional fuels. In order to acquire market share, sustainable fuels must exhibit their cost-competitiveness and performance advantages. This necessitates ongoing innovation, the utilization of economies of scale, and beneficial policies to equalize the opportunities.

Increasing the scale of sustainable fuel production and distribution presents substantial obstacles. The establishment of sufficient infrastructure, encompassing manufacturing facilities, transportation networks, and storage capacities, necessitates significant financial commitments and meticulous strategic planning. Furthermore, it is crucial to guarantee the sustainability and traceability of feedstocks and production processes in order to preserve the environmental integrity of these fuels. To meet the increasing demand for sustainable fuels and guarantee their widespread availability, it is crucial to overcome these supply chain issues.

Although considerable strides have been achieved in the development of sustainable fuel technologies, additional improvements are required to enhance their efficiency, decrease expenses, and further optimize their performance. The research and development endeavors should prioritize the resolution of technical obstacles, the optimization of production processes, and the fabrication of novel solutions. In order to overcome economic obstacles, such as exorbitant initial expenses and unpredictable investment returns, it is necessary to implement supportive policies, financial incentives, and risk-sharing frameworks. The achievement of cost parity with conventional jet fuel and the scaling up of production are key challenges in the development of sustainable aviation fuels (SAFs) [19].

However promising, the future market for sustainable fuels encounters several obstacles. To surmount these obstacles, it is imperative that governments, industries, and investors collaborate tirelessly. Sustainability fuels can significantly contribute to the attainment of a cleaner and more sustainable energy future by effectively addressing the competition with current fuels, guaranteeing scalability and supply chain resilience, and surmounting technological and economic obstacles.

9.3 Utilization of Sustainable Energy Carriers

9.3.1 Hydrogen as a Key Player in Future Energy Systems

The high energy density and zero-carbon emissions at the point of use of hydrogen are positioning it as a significant contributor to the shift towards a sustainable energy future. The versatility and potential applications of this technology in several sectors such as industry, transportation, and power generation make it an essential element of the worldwide decarbonization initiatives.

Ammonia production and oil refining are among the industrial processes that already employ hydrogen. Nevertheless, its potential surpasses these conventional uses by a significant margin. Green hydrogen, derived from sustainable energy sources, has the potential to reduce carbon emissions in industries that are difficult to decarbonize, such as steelmaking and chemical extraction, where direct conversion to electricity is difficult. The HYBRIT project in Sweden is investigating the use of green hydrogen as a substitute for coal in steel manufacturing, with the goal of attaining steel that is free from fossil fuels by 2026 [20].

Considerable progress is being made in the generation of green hydrogen, mainly by electrolysis fueled by renewable energy. Energy-efficient and cost-competitive electrolyzer technologies are enhancing the feasibility of green hydrogen. The primary objective of research and development is to enhance the efficiency of electrolyzers, minimize their cost, and advance the innovation of production techniques, including photoelectrochemical water splitting and high-temperature electrolysis. Industrial enterprises such as ITM Power are leading the way in the development and implementation of large-scale electrolyzers to facilitate the generation of environmentally friendly hydrogen for diverse uses [21].

The effective implementation of hydrogen requires the establishment of a strong infrastructure, encompassing manufacturing plants, transportation systems, and storage capacities. The establishment of hydrogen pipelines, the optimization of current natural gas pipelines, and the advancement of hydrogen storage technologies, such as compressed gas and liquid hydrogen, are essential for enabling the efficient distribution and use of hydrogen. Initiatives such as the HyNet North West project in the UK, which seeks to establish a hydrogen infrastructure network, are validating the viability of establishing a hydrogen-based economy [22].

The potential of hydrogen to reduce carbon emissions in several industries and contribute to a sustainable energy future positions it as a crucial player in the worldwide energy transition. Significant progress in the production of environmentally friendly hydrogen, together with the establishment of a strong hydrogen infrastructure, is facilitating the emergence of a hydrogen-based economy. In the pursuit of net-zero emissions goals, hydrogen is positioning itself as a crucial facilitator of a more environmentally friendly and enduring energy landscape.

9.3.2 Ammonia as an Energy Carrier and Fuel

Ammonia, a nitrogen–hydrogen compound, is increasingly acknowledged as a highly promising energy carrier and fuel because of its exceptional energy density, convenient storage and transportation capabilities, and potential for application in low-carbon technologies. Given its adaptability and capacity to reduce carbon emissions in several industries, especially in marine transportation and power production, it is a compelling choice for achieving a sustainable energy future.

The exceptional energy density and liquid state of ammonia at normal temperatures render it highly suitable for marine transportation, so presenting a promising substitute for heavy fuel oil and other fossil fuels. Multiple initiatives are now implemented to advance the development of ships powered by ammonia, so showcasing the practicality of this technology for reducing carbon emissions in the maritime sector. Furthermore, ammonia has the capability to be utilized in power production, either by direct use in combustion turbines or indirectly through fuel cells, so offering a versatile and potentially environmentally friendly energy medium. MAN Energy Solutions, a prominent engine manufacturer, is currently engaged in the development of ammonia-fueled engines specifically designed for marine applications. This initiative highlights the considerable potential of this fuel to revolutionize the shipping industry [23].

Despite the numerous benefits of ammonia, its safe and efficient transportation and storage present specific difficulties. Given its toxicity and corrosive nature, ammonia necessitates meticulous management and specialized infrastructure. Ensuring the safe and economical transportation and storage of ammonia is essential to enable its widespread use as an energy carrier. The primary objective of research and development is to tackle these issues by developing sophisticated materials and technologies for the containment and transportation of ammonia.

The potential function of ammonia extends beyond its primary applications as an energy carrier and fuel. Additionally, it can function as a valuable raw material for several industrial operations, including the manufacturing of fertilizers and the synthesis of complex chemicals. Within a circular economy, which involves the reuse and recycling of resources, ammonia can have a vital function in completing the cycle and reducing environmental waste. For example, the carbon dioxide that is captured can be used in the manufacturing of green ammonia, so further decreasing its carbon footprint and making a valuable contribution to a more environmentally efficient energy system.

Given its adaptability and capacity for low-carbon uses, ammonia is becoming a significant energy carrier and fuel in the shift towards a sustainable energy future. Its application in maritime transportation and electricity production, together with its contribution to a circular economy, renders it a compelling choice for decarbonization endeavors. While obstacles persist in terms of transportation and storage, continuous research and development are facilitating a broader acceptance of ammonia as a significant participant in the global energy sector.

9.3.3 *Methanol in the Transport and Power Sectors*

Methanol, a simple alcohol derived from various sources including natural gas, coal, and biomass, shows great potential as a sustainable fuel in the transportation and power industries. Given its high energy density, ease of handling, and potential for lower emissions than traditional fossil fuels, it emerges as a compelling option in the pursuit of decarbonization.

The compatibility of methanol with current internal combustion engine technology offers a rather smooth route for its implementation in the transportation industry. By making slight adjustments, traditional gasoline engines can operate with high efficiency on methanol, providing a readily accessible avenue for decreasing emissions in the short term. Furthermore, the high octane rating and clean-burning characteristics of methanol can result in enhanced engine performance and decreased air pollution. China and Sweden are among the countries currently implementing programs to encourage the use of methanol as a fuel for transportation, indicating its capacity for extensive acceptance [24].

Furthermore, apart from its application in internal combustion engines, methanol shows potential as a fuel for fuel cells, providing a means to achieve zero-emission power production in both automobiles and stationary power plants. The direct conversion of methanol into electricity by direct methanol fuel cells (DMFCs) offers a compact and highly efficient power source suitable for a wide range of applications. Within the transportation industry, DMFCs are being investigated for application in light-duty vehicles, buses, and aircraft. Methanol fuel cells in the power industry has the capability to supply environmentally friendly and dependable electricity to isolated or off-grid areas, as well as serve as a backup power source for essential infrastructure. Industry players such as Blue World Technologies are actively engaged in the development and commercialization of methanol fuel cell solutions for a wide range of applications, so underscoring the increasing fascination with this particular technology [25].

Due to its adaptability and capacity to significantly lower emissions, methanol is establishing a specialized position in the transportation and power industries. The utilization of this substance as a fuel for internal combustion engines presents a viable interim resolution for reducing carbon emissions, while its implementation in fuel cells presents a route towards achieving power generation without any emissions. With ongoing technical progress and the implementation of supportive regulatory measures, methanol is positioned to assume a progressively important position in the shift towards a sustainable energy future.

9.3.4 DME in Transportation and Power Generation

DME, a synthetic fuel derived from methanol or biomass, is increasingly being acknowledged as a feasible substitute for diesel fuel in the fields of transportation and power generation. The combination of its clean-burning characteristics, high cetane number, and ease of handling renders it a compelling choice for mitigating carbon emissions and enhancing air quality.

The physical and chemical characteristics of DME closely mimic those of diesel fuel, rendering it a viable substitute for diesel engines with modest adjustments. Efficient combustion is achieved due to its high cetane number, which effectively reduces emissions of particulate matter and nitrogen oxide. Multiple pilot projects and demonstrations have effectively demonstrated the applicability of DME in buses, trucks, and marine vessels, so emphasizing its potential for extensive implementation in the transportation industry. For instance, the Volvo Group has undertaken thorough research and development on DME as a fuel for heavy-duty vehicles, showcasing its capacity to attain substantial reductions in emissions [26].

DME can be derived from sustainable sources, such as biomass or captured carbon dioxide, resulting in a substantial decrease in greenhouse gas emissions in comparison with traditional crude diesel fuel. The cleaner combustion characteristics of this fuel also enhance air quality, especially in metropolitan regions where diesel emissions are a significant issue. Utilizing DME as a propulsion fuel can significantly contribute to the attainment of decarbonization objectives and the reduction of the environmental consequences associated with the transport industry.

The effective implementation of DME depends on the establishment of large-scale manufacturing facilities and the necessary infrastructure. To attain cost competitiveness, the production of DME from renewable sources necessitates additional technical progress and the realization of economies of scale. Furthermore, the development of distribution networks and refueling stations is essential to enable the deployment of DME in the transportation industry. Numerous enterprises, including Oberon Fuels, are actively engaged in the development of DME production technologies and the advancement of its commercialization [27].

The potential of DME to substitute diesel fuel in transportation and power generation presents a promising avenue for mitigating carbon emissions and enhancing air quality. Enhancements in manufacturing technologies, together with the establishment of necessary infrastructure, are essential to fully harness the potential of DME as a sustainable fuel. In the ongoing global shift towards a more environmentally friendly energy future, DME is well-positioned to have a substantial impact on reducing carbon emissions in the transportation industry and encouraging a more sustainable energy environment.

9.3.5 *Integration of Sustainable Carriers into Existing Energy Systems*

The shift towards sustainable energy carriers necessitates the advancement of novel technologies and infrastructure, as well as the incorporation of these carriers into current energy systems. This entails retrofitting current infrastructure, integrating novel energy carriers with conventional fuels, and guaranteeing the sustained economic feasibility of widespread implementation.

The integration of hydrogen with natural gas in current pipelines and infrastructure presents a highly promising approach to reduce carbon emissions in the gas industry. By blending hydrogen with natural gas at different proportions, it is possible to decrease the carbon intensity of the fuel mix without necessitating substantial alterations to current infrastructure. Multiple ongoing pilot projects and demonstrations are being conducted to assess the viability and safety of hydrogen blending, yielding encouraging outcomes. The UK-based HyDeploy project is effectively incorporating up to 20% hydrogen into the natural gas network, so showcasing the potential of this strategy in mitigating emissions within the gas industry [28].

To expedite the shift to a cleaner energy system, it is possible to repurpose current infrastructure, such as pipelines and power plants, for sustainable energy carriers. One example is the modification of current natural gas pipelines to facilitate the transportation of hydrogen or other environmentally friendly gases. Similarly, coal-fired power plants can be upgraded to co-fire with biomass or adopt carbon capture and storage technologies. By retrofitting current infrastructure, it is possible to utilize existing assets, minimize the requirement for new construction, and expedite the implementation of sustainable energy solutions.

An essential factor to consider is the long-term economic viability of widespread implementation of sustainable energy carriers. Despite the substantial initial expenses associated with adopting new technologies and infrastructure, the long-term advantages, such as decreased emissions, better air quality, and increased energy security, can surpass the initial expenditures. Implementing supportive measures, such as carbon pricing and incentives for the integration of sustainable fuels, can significantly improve the economic feasibility of these solutions. Furthermore, the ongoing progress in technology and the realization of economies of scale are anticipated to gradually reduce the expenses associated with sustainable energy carriers, so enhancing their competitiveness in comparison with traditional fuels.

The incorporation of sustainable energy carriers into current energy systems is a multifaceted yet essential endeavor. Essential measures for achieving a cleaner and more sustainable energy future include combining hydrogen with natural gas, upgrading current infrastructure, and guaranteeing the long-term economic viability of these solutions.

9.4 Concluding Remarks

This book outlines the comprehensive trajectory of sustainable energy carriers, encompassing molecular design, synthesis methodologies, storage frameworks, transport logistics, techno-economic feasibility, and environmental impact, within the overarching evolution of energy systems. We demonstrated how hydrogen, ammonia, methanol, dimethyl ether, formic acid, and other liquid/solid carriers can bridge the divide between intermittent renewables and end-use sectors requiring dense, dispatchable energy. Three constants emerged throughout the chapters: (1) production pathways must achieve decarbonization via renewable electricity, carbon capture, and circular feedstocks; (2) storage and transportation decisions depend on safety, cost, and energy-density trade-offs that vary by carrier; and (3) systems integration—not merely molecule selection—ultimately dictates climate impact and market adoption.

Our comparative evaluations indicated that no single carrier excels across all metrics. Hydrogen demonstrates superior conversion efficiency but suffers from low volumetric density; ammonia provides compatibility for global shipping but introduces NOx emissions and safety issues; methanol and DME utilize current petrochemical and LPG infrastructure yet rely on low-carbon CO_2 and H_2 sources; formic acid and liquid organic hydrogen carriers (LOHCs) facilitate handling under ambient conditions but necessitate catalytic improvements for effective release. The life-cycle analyses indicated that the upstream electricity mix, plant scale, and by-product management frequently eclipse marginal improvements in reactor efficiency. Simultaneously, economic assessments highlighted the importance of electrolyzer capital expenditures, carbon pricing, and learning curves, indicating that policy support should be adaptive and grounded in empirical evidence.

In the future, decentralized grids, digitalized operations, hybrid renewable-CCUS systems, and strong regulatory frameworks will determine which carriers thrive. The establishment of markets for hydrogen and ammonia, funding mechanisms for pioneering plants, and the development of standardized safety and quality protocols are urgent priorities. Sectoral analyses—heavy transport, marine fuels, industrial heat, and power balancing—underscore the areas where each carrier currently provides the greatest value and where research, development, and demonstration should concentrate to unveil future possibilities. In conclusion, sustainable energy carriers serve as facilitating connections within a broader net-zero framework. Their success will depend on synchronized advancements in catalysis, electrochemistry, infrastructure modifications, digital supply chain management, and conducive governance. This book seeks to provide researchers, industry leaders, and policymakers with a structured framework for effectively deploying appropriate molecules in suitable locations, thereby facilitating a fair and robust transition to a sustainable energy future.

References

1. Orsini L, Kessler S, Wei J, Field H (2019) How the Brooklyn microgrid and TransActive grid are paving the way to next-gen energy markets. In: The energy internet. Elsevier, pp. 223–239. https://doi.org/10.1016/B978-0-08-102207-8.00010-2
2. Zhang C, Yang T, Wang Y (2021) Peer-to-peer energy trading in a microgrid based on iterative double auction and blockchain. Sustain Energy Grids Netw 27:100524. https://doi.org/10.1016/j.segan.2021.100524
3. Quitzow L, Canzler W, Grundmann P, Leibenath M, Moss T, Rave T (2016) The German Energiewende—what's happening? Introducing the special issue. Util Policy 41:163–171. https://doi.org/10.1016/j.jup.2016.03.002
4. Ren D, Lu L, Hua R, Zhu G, Liu X, Mao Y et al (2023) Challenges and opportunities of practical sulfide-based all-solid-state batteries. E-Transportation 18:100272. https://doi.org/10.1016/j.etran.2023.100272
5. Hyundai NEXO (2019) SUV passes crash testing in US, showcase in China. Fuel Cells Bull 2019:2–2. https://doi.org/10.1016/s1464-2859(19)30357-8
6. Rangarajan A, Foley S, Trück S (2023) Assessing the impact of battery storage on Australian electricity markets. Energy Econ 120:106601. https://doi.org/10.1016/j.eneco.2023.106601

7. Stopps H, Touchie MF (2021) Residential smart thermostat use: an exploration of thermostat programming, environmental attitudes, and the influence of smart controls on energy savings. Energy Build 238:110834. https://doi.org/10.1016/j.enbuild.2021.110834
8. Siemens (2024). Gridscale X meter data management https://www.siemens.com/global/en/products/energy/grid-software/meter-data-management.html. Accessed 9 Sept 2024
9. Corbet S, Goodell JW (2022) The reputational contagion effects of ransomware attacks. Financ Res Lett 47:102715. https://doi.org/10.1016/j.frl.2022.102715
10. Wu H, West SR (2024) Co-optimisation of wind and solar energy and intermittency for renewable generator site selection. Heliyon 10:e26891. https://doi.org/10.1016/j.heliyon.2024.e26891
11. Melián-Martel N, del Río-Gamero B, Schallenberg-Rodríguez J (2021) Water cycle driven only by wind energy surplus: towards 100% renewable energy islands. Desalination 515:115216. https://doi.org/10.1016/j.desal.2021.115216
12. Bisotti F, Hoff KA, Mathisen A, Hovland J (2024) Direct air capture (DAC) deployment: a review of the industrial deployment. Chem Eng Sci 283:119416. https://doi.org/10.1016/j.ces.2023.119416
13. Mantripragada HC, Zhai H, Rubin ES (2019) Boundary Dam or Petra Nova—which is a better model for CCS energy supply? Int J Greenh Gas Control 82:59–68. https://doi.org/10.1016/j.ijggc.2019.01.004
14. IEA (2021) Ammonia technology roadmap. IEA. https://doi.org/10.1787/f6daa4a0-en
15. (2020) Air products to build $5bn green hydrogen facility in Saudi Arabia. Fuel Cells Bull 2020:11–11. https://doi.org/10.1016/s1464-2859(20)30357-6
16. EC (2022) EU emissions trading system (EU ETS). Eur Comm. https://climate.ec.europa.eu/eu-action/eu-emissions-trading-system-eu-ets_en. Accessed 30 May 2023
17. Kabir Z, Yusuf MA, Khan I (2021) An overview of policy framework and measures promoting bioenergy usage in the EU, the United States, and Canada. In: Bioenergy resources and technologies. Elsevier, pp 383–421. https://doi.org/10.1016/B978-0-12-822525-7.00015-9
18. Hy24 (2024) Scaling up the hydrogen economy. https://www.hy24partners.com/. Accessed 9 Sept 2024
19. Watson MJ, Machado PG, da Silva AV, Saltar Y, Ribeiro CO, Nascimento CAO et al (2024) Sustainable aviation fuel technologies, costs, emissions, policies, and markets: a critical review. J Clean Prod 449:141472. https://doi.org/10.1016/j.jclepro.2024.141472
20. Karakaya E, Nuur C, Assbring L (2018) Potential transitions in the iron and steel industry in Sweden: towards a hydrogen-based future? J Clean Prod 195:651–663. https://doi.org/10.1016/j.jclepro.2018.05.142
21. ITM (2024) New full scope 5MW containerised electrolyser. ITM Power. https://itm-power.com/index.html. Accessed 9 Sept 2024
22. Sovacool BK, Iskandarova M, Geels FW (2023) "Bigger than government": exploring the social construction and contestation of net-zero industrial megaprojects in England. Technol Forecast Soc Change 188:122332. https://doi.org/10.1016/j.techfore.2023.122332
23. MAN (2024) Shaping the future for hydrogen | MAN Energy Solutions. https://www.man-es.com/company/about-us/hydrogen/2022/08/03/storengy-chooses-man-energy-solutions-for-methanation-reactor-to-produce-syngas-at-a-french-wastewater-treatment-plant. Accessed 4 Nov 2023
24. Svanberg M, Ellis J, Lundgren J, Landälv I (2018) Renewable methanol as a fuel for the shipping industry. Renew Sustain Energy Rev 94:1217–1228. https://doi.org/10.1016/j.rser.2018.06.058
25. (2018) Blue World to produce methanol fuel cells for the mobility market. Fuel Cells Bull 2018:11–12. https://doi.org/10.1016/s1464-2859(18)30423-1
26. Peinado C, Liuzzi D, Sluijter SN, Skorikova G, Boon J, Guffanti S et al (2024) Review and perspective: next generation DME synthesis technologies for the energy transition. Chem Eng J 479:147494. https://doi.org/10.1016/j.cej.2023.147494
27. Oberon (2024) Fuel better renewable methanol, DME and hydrogen. Oberon Fuels. https://www.oberonfuels.com/. Accessed 9 Sept 2024

References

28. (2018) HyDeploy P2G project aims to reduce home CO_2 emissions in UK. Fuel Cells Bull 2018:11–11. https://doi.org/10.1016/S1464-2859(18)30091-9

Open Access This chapter is licensed under the terms of the Creative Commons Attribution 4.0 International License (http://creativecommons.org/licenses/by/4.0/), which permits use, sharing, adaptation, distribution and reproduction in any medium or format, as long as you give appropriate credit to the original author(s) and the source, provide a link to the Creative Commons license and indicate if changes were made.

The images or other third party material in this chapter are included in the chapter's Creative Commons license, unless indicated otherwise in a credit line to the material. If material is not included in the chapter's Creative Commons license and your intended use is not permitted by statutory regulation or exceeds the permitted use, you will need to obtain permission directly from the copyright holder.

The manufacturer's authorised representative in the EU is Springer Nature Customer Service Centre GmbH, Europaplatz 3, 69115 Heidelberg, Germany. If you have any concerns regarding our products, please contact ProductSafety@springernature.com

Printed and bound by CPI Group (UK) Ltd, Croydon, CR0 4YY

27/03/2026

02080183-0001